Renewable Energy

Editor: Tracy Biram

Volume 344

Independence Educational Publishers

First published by Independence Educational Publishers

The Studio, High Green

Great Shelford

Cambridge CB22 5EG

England

© Independence 2019

ISBN-13: 978 1 86168 800 2

Printed in Great Britain

Zenith Print Group

Contents

Introduction

RENEWABLE ENERGY is Volume 344 in the **ISSUES** series. The aim of the series is to offer current, diverse information about important issues in our world, from a UK perspective.

ABOUT RENEWABLE ENERGY

Renewable energy sources generate about a third of the UK's electricity. We are all aware that we need to help the environment, and by using renewable energy is one way we can help. But just how do we decide which is the right one for us? How can we benefit from renewable energy? This book discusses the ways in which we can use energy to help the environment.

OUR SOURCES

Titles in the **ISSUES** series are designed to function as educational resource books, providing a balanced overview of a specific subject.

The information in our books is comprised of facts, articles and opinions from many different sources, including:

⇨ Newspaper reports and opinion pieces

⇨ Website factsheets

⇨ Magazine and journal articles

⇨ Statistics and surveys

⇨ Government reports

⇨ Literature from special interest groups.

A NOTE ON CRITICAL EVALUATION

Because the information reprinted here is from a number of different sources, readers should bear in mind the origin of the text and whether the source is likely to have a particular bias when presenting information (or when conducting their research). It is hoped that, as you read about the many aspects of the issues explored in this book, you will critically evaluate the information presented.

It is important that you decide whether you are being presented with facts or opinions. Does the writer give a biased or unbiased report? If an opinion is being expressed, do you agree with the writer? Is there potential bias to the 'facts' or statistics behind an article?

ASSIGNMENTS

In the back of this book, you will find a selection of assignments designed to help you engage with the articles you have been reading and to explore your own opinions. Some tasks will take longer than others and there is a mixture of design, writing and research-based activities that you can complete alone or in a group.

Useful weblinks

www.blueandgreentomorrow.com

www.edfenergy.com

www.energysavingtrust.org.uk

www.energyvoice.com

www.euronews.com

www.fullfact.org

www.greenchildmagazine.com

www.greenmatch.co.uk

www.greenpeace.org.uk

www.imperial.ac.uk

www.independent.co.uk

www.inews.co.uk

www.innovateuk.blog.gov.uk

www.labmate-online.com

www.lse.ac.uk/GranthamInstitute

www.ofgem.gov.uk

www.openaccessgovernment.org

www.renewableenergyhub.co.uk

www.theconversation.com

www.theecologist.org

www.theguardian.com

FURTHER RESEARCH

At the end of each article we have listed its source and a website that you can visit if you would like to conduct your own research. Please remember to critically evaluate any sources that you consult and consider whether the information you are viewing is accurate and unbiased.

Types and alternative sources of renewable energy

By Laura Davies

What is a renewable energy source?

A renewable energy source means energy that is sustainable – something that can't run out, or is endless, like the Sun. When you hear the term 'alternative energy' it's usually referring to renewable energy sources too. It means sources of energy that are alternative to the most commonly used non-sustainable sources like coal.

What isn't a renewable energy source?

Nuclear power is an alternative to traditional fossil fuels such as coal, oil and gas. It creates fewer airborne pollutants too but it is nonetheless a finite resource. Despite the fact that generating electricity from nuclear energy is a very different process to burning fossil fuels, nuclear power is not a renewable energy source.

As we rely on fossil energy sources we need to make them as efficient as possible. Although cleaner, more efficient technologies that reduce the environmental impact of fossil fuels are desirable, they are not renewable energy sources.

Burning wood instead of coal is slightly more complex. On the one hand, wood is a renewable resource – provided it comes from sustainably managed forests. Wood pellets and compressed briquettes are made from by-products of the wood processing industry and so arguably reduce waste. Compressed biomass fuels produce more energy than logs too. On the other hand, burning wood (whether it be raw timber or processed waste) releases particles into our atmosphere. Burning wood always results in deforestation and the reduction of natural habitats, so although it can be a renewable energy source, it's not what we would call 'alternative'.

Types of renewable energy

Alternative or renewable energy comes from natural processes that (unlike those listed above) can reliably produce cheap energy with minimal impact to the environment. The most popular renewable energy sources currently are:

Solar energy

Sunlight is one of our planet's most abundant and freely available energy resources. The amount of solar energy that reaches the Earth's surface in one hour is more than the planet's total energy requirements for a whole year. Although it sounds like a perfect renewable energy source, the amount of solar energy we can use varies according to the time of day and the season of the year as well as geographical location. In the UK, solar energy is an increasingly popular way to supplement your energy usage.

Wind energy

Wind is a plentiful source of clean energy. Wind farms are an increasingly familiar sight in the UK with wind power

Renewable energy facts

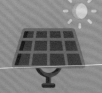

Solar PV could account for 5% of global demand by 2020 and up to 9% by 2030

Price Waterhouse Cooper predicts that Africa could run on 100% renewable energy by 2050

By the year 2050, our energy needs can be met by 95% renewable energy

Over the last four decades, the price of solar PV panels has declined 99%

A US study showed that renewable energy creates three times more jobs than fossil fuels

Investment in renewable energy has surpassed fossil fuel investment. The global renewable energy market is now worth over $250 billion

Source: EDF Energy

making an ever-increasing contribution to the National Grid. To harness electricity from wind energy, turbines are used to drive generators which then feed electricity into the National Grid. Although domestic or 'off-grid' generation systems are available, not every property is suitable for a domestic wind turbine.

Hydro energy

As a renewable energy resource, hydro power is one of the most commercially developed. By building a dam or barrier, a large reservoir can be used to create a controlled flow of water that will drive a turbine, generating electricity. This energy source can often be more reliable than solar or wind power (especially if it's tidal rather than river) and also allows electricity to be stored for use when demand reaches a peak. Like wind energy, in certain situations hydro can be more viable as a commercial energy source (dependant on type and compared to other sources of energy) but depending very much on the type of property, it can be used for domestic, 'off-grid' generation.

Tidal energy

This is another form of hydro energy that uses twice-daily tidal currents to drive turbine generators. Although tidal flow unlike some other hydro energy sources isn't constant, it is highly predictable and can therefore compensate for the periods when the tide current is low.

Geothermal energy

By harnessing the natural heat below the Earth's surface, geothermal energy can be used to heat homes directly or to generate electricity. Although it harnesses a power directly below our feet, geothermal energy is of negligible importance in the UK compared to countries such as Iceland, where geothermal heat is much more freely available.

Biomass energy

This is the conversion of solid fuel made from plant materials into electricity. Although, fundamentally, biomass involves burning organic materials to produce electricity, this is not burning wood, and nowadays this is a much cleaner, more energy-efficient process. By converting agricultural, industrial and domestic waste into solid, liquid and gas fuel, biomass generates power at a much lower economical and environmental cost.

The cost of renewable energy

In 2016, global renewable energy capacity grew by a record amount while its cost fell considerably. This improvement was largely due to a drop in the cost of both solar and wind energy.

161GW of renewable energy capacity was installed worldwide in 2016 – a 10% rise on the preceding year and a new record as reported by REN21, a global renewable energy policy network covering 155 nations and 96% of the world's population.

New solar installation provided the biggest boost – half of all new capacity. Wind power accounted for a further third with hydropower accounting for 15%. New solar capacity exceeded all other electricity producing technology for the first time ever.

Although traditional fossil fuels such as oil and gas still provide 80% of global electricity output, renewables are rising fast. Christina Figueres, the former UN climate chief said: 'The economic case for renewables as the backbone of our global energy system is increasingly clear and proven. Offering greater bang-for-buck, renewables are quite simply the cheapest way to generate energy in an ever-growing number of countries.

Renewable energy and your home

The advantages of using renewable energy in a domestic setting are persuasive:

⇨ Cut your electricity bills: Once you've paid for the costs of installing a renewable energy system, you can become less reliant on the National Grid and your energy bills can be reduced.

⇨ Get paid for the electricity you generate: The UK Government's Feed-in Tariff pays you for the electricity you generate, even if you use it.

⇨ Sell electricity back to the grid: If you are generating enough energy to export an excess back into the National Grid, you can receive an additional payment from the Feed-in Tariff scheme.

⇨ Reduce your carbon footprint: Green, renewable sources of energy don't release carbon dioxide or other harmful pollutants into the atmosphere. According to the Energy Saving Trust's Solar panels page, a typical solar PV system could save around 1.5–2 tonnes of carbon per year.

New renewable energy sources

To reduce our dependence on traditional fossil fuels, scientists are increasingly looking for new, renewable energy sources. Some of them might surprise you.

21 December 2017

www.edfenergy.com

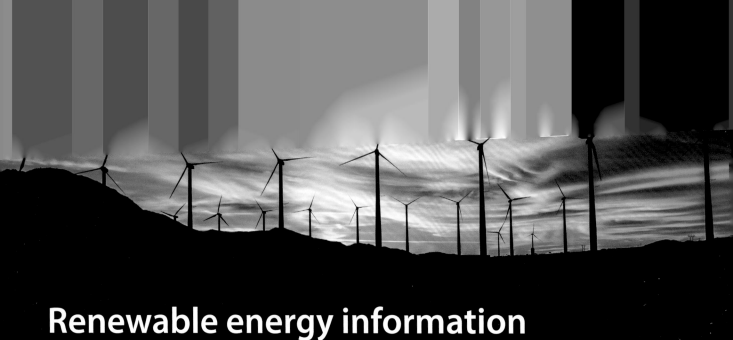

Renewable energy information

An introduction to renewable energy

If we are going to reverse the effect of climate change then we need to begin using renewable energy sources to heat our homes, light our offices, power our machines, and drive our cars. A renewable energy is one which is replenished naturally and constantly such as sunlight, wind, tidal power, biomass, and thermal energy. There are some good reasons for moving towards more sustainable and green energy sources including:

⇨ Counteracting the effects of climate change and reducing our reliance on fossil fuels such as coal and oil.

⇨ Decreasing reliance on other countries for energy needs, in other words creating energy security.

⇨ Finding cheaper and cleaner ways of providing electricity and heating our homes.

History of renewable energy

We have always been trying to find new ways to use our environment to produce power. Wood burning was one of the first biomass renewable energy projects and has been around since man discovered fire. Wind has been used to power ships and drive windmills for grinding corn since the Egyptian times and the first experiments in solar power began way back in the late 19th century.

Many consider that the use of coal and then oil took attention away from further developing renewable energy until we began to realise the effect on the environment and the problems of sustaining an economy based on the use of fossil fuels. Gas, coal and oil will run out at some point and they won't be replenished for several million years.

For the last 20–30 years, research and development into finding more sustainable ways for providing our energy needs has been growing rapidly. World energy consumption just four years ago was 80% dependent on fossil fuels, a situation that cannot be maintained indefinitely. From the development of solar panels and wind power, to electric-powered cars, the challenge to lower our carbon footprint and find cleaner energy sources has been accepted and is now driving forward at pace.

Mainstream renewable energy sources

If we are going to replace our dependence on fossil fuels then we will need to draw energy from a number of diverse sources including the Sun, wind, the sea, rivers, and natural environment.

Wind power

Drive across the UK landscape 30 years ago and you might have come across a wind turbine or two here and there. Nowadays, we are used to seeing large wind farms both on land and at sea, those huge white propellers turning languidly in the breeze. At the beginning of 2014, there were some 5,276 wind turbines in the UK with a total capacity of ten gigawatts, roughly equally distributed between land-based farms and offshore ones.

Of all the renewable technologies, wind turbines are the ones that garner the most controversy. Because of their size they are often seen as a blot on the landscape and there are many who see investment in this technology as unwise because it supposedly only covers a small proportion of the UK's energy needs.

The truth is that investment in wind power is set to increase despite recent cuts in government subsidies.

Solar power

The conversion of solar energy to power has been in use for a long time, since the days we discovered greenhouses and how a magnifying glass can burn paper. Nowadays, our development of solar technologies such as photovoltaic cells, solar panels and solar thermal collectors means that this renewable energy is at the forefront of our fight to

find better and more sustainable ways to provide our electricity and heating needs.

Solar panels can be used on our roofs and in huge farms, to produce electricity that can be fed into the grid. It's one of the true renewable energies where domestic and commercial concerns have an equal stake.

Solar thermal is the technology now being used to heat water and is utilised by many businesses. While it is widely found throughout continental Europe, in the UK it has been under used to date but is one of the most effective and cheapest ways of providing heating for our buildings.

Biomass energy

Biomass is plant material that can be either burned as fuel or converted to biofuels, to drive machinery, even cars.

The biomass material most used for fuel is wood, which can

be found in low carbon boilers that produce heating and hot water for homes. These systems are increasingly replacing coal, oil and gas heating systems for both businesses and domestic houses.

Biofuels can be made from a variety of different organisms producing solids, liquids and gases that can be mass produced to replace fossil fuels such as petrol. Still in its early days, biofuel development could hold the key to powering our cars and other machines, though there is some doubt as to whether they address the concerns of global warming.

Hydropower

One of the burgeoning renewable industries of the last 20 years is the use of hydropower. We have long since harnessed the rushing water of lakes up on high land with

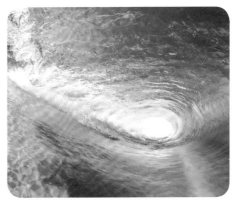

power stations such as Dinorwig in North Wales. Now we are beginning to develop smaller systems that can make a real difference to our energy production.

More recent innovations are attempts to harness the power of our coastline with wave energy machines such as the Oceanus 2 project in the South East of England.

Geothermal energy

The heat held in the earth is a potent source of energy, if it can be efficiently harnessed. Hot springs have been used by different civilisations for centuries but actually developing the technology to produce electricity on a national scale has

proved difficult. In the UK, it has been explored as a possible energy alternative although it has been restricted so far to shallow geothermal sites that produce a limited amount of power for the National Grid. While there are geothermal hotspots in Cornwall, East Yorkshire and the Lake District, the technology and finances have not yet

been provided to yield this possibly useful power source. Renewable energy lies at the heart of our future. The developments are beginning to come thick and fast and advocates hope that we soon reach the tipping point where our dependence on fossil fuels is outweighed by our own, natural energy production, allowing us to be sustainable and, more importantly, self-sufficient.

www.renewableenergyhub.co.uk

Renewable energy set to be cheaper than fossil fuels by 2020, according to new report

'Turning to renewables for new power generation is not simply an environmentally conscious decision, it is now – overwhelmingly – a smart economic one.'

By Josh Gabbatiss, Science Correspondent

Renewable energy will be cheaper than fossil fuels in two years, according to a new report. Experts predict that investment in green infrastructure projects will lead to decreases in the cost of energy for consumers. Continuous technological improvements have led to a rapid fall in the cost of renewable energy in recent years, meaning some forms can already comfortably compete with fossil fuels.

The report suggests this trend will continue, and that by 2020 'all the renewable power generation technologies that are now in commercial use are expected to fall within the fossil fuel-fired cost range'. Of those technologies, most will either be at the lower end of the cost range or actually undercutting fossil fuels.

'This new dynamic signals a significant shift in the energy paradigm,' said Adnan Amin, director-general of the International Renewable Energy Agency (IREA), which published the report.

'Turning to renewables for new power generation is not simply an environmentally conscious decision, it is now – overwhelmingly – a smart economic one.'

The report looked specifically at the relative cost of new energy projects being commissioned. As renewable energy becomes cheaper, consumers will benefit from investment in green infrastructure.

'If the stuff you're building to generate electricity costs less, the end effect of that is having to pay less for the electricity that comes from it,' Jonathan Marshall, energy analyst at the Energy and Climate Intelligence Unit (ECIU) told *The Independent*.

'The cheaper you install it, the better it is for everyone.'

The current cost for fossil fuel power generation ranges from around 4p to 12p per kilowatt hour across G20 countries. By 2020, IREA predicted renewables will cost between 2p and 7p, with the best onshore wind and solar photovoltaic projects expected to deliver electricity by 2p or less next year.

Other methods of producing renewable energy, such as offshore wind farms and solar thermal energy, are not yet as competitive as fossil fuels.

However, the results of recent renewable power auctions for projects to be commissioned in the coming years suggest these forms too are due to drop in price.

Auctions provide a useful means of predicting the future cost of electricity.

'These cost declines across technologies are unprecedented and representative of the degree to which renewable energy is disrupting the global energy system,' said Mr Amin.

The new report comes after 2017 was declared the UK's 'greenest year ever' by WWF, when data from the National Grid revealed 13 different renewable energy records had been broken.

However, current UK policy may hamper the development of renewable energy capacity.

'Under current policy, the UK is at risk of being left behind as other countries take full advantage of the relentless fall in the cost of renewable energy,' said Mr Marshall.

Notably, the subsidy ban for new onshore wind farms has been singled out, with the ECIU predicting it could add £1 billion onto energy bills over five years.

'If the Government is serious about achieving the lowest cost electricity in Europe, the ban on onshore wind has to be first in the firing line,' said Mr Marshall.

'Until this happens – and all low-carbon electricity sources are allowed to compete on equal footing – the gap between the cost of electricity in the UK and elsewhere will prevail; to the ire of politicians, businesses and household bill payers.'

A spokesperson from the Department for Business, Energy & Industrial Strategy said the Government could still support onshore wind where there is local support, such as on the Scottish islands.

'We are pleased to see that established technologies, such as onshore wind and solar, are driving costs down for consumers,' they said.

'If this continues, and they have local support, they may play a significant role in the energy mix in future.

'Since 2010, the UK has invested more than £52 billion in renewable energy and in October, we confirmed that up to £557 million would be made available for future clean power auctions.'

15 January 2018

Eight awesome facts about renewable energy

By Anum Yoon

Green energy is a simple concept. It's using renewable energy sources like wind, solar, water or geothermal to generate electricity instead of traditional sources. Some of the most beneficial facts about renewable energy are that it produces fewer pollutants, runs cheaper, and can also create jobs

The only downside of green energy is that it pushes fossil fuels out, and people are often reluctant to change… especially when money is involved.

We're living in a wonderful time where young people are fuelling social change and helping to keep the environment safe. This generation can grow up learning about expecting clean energy, as long as we start teaching them early.

Eight facts about renewable energy

Here are 8 facts about renewable energy you can share with the young people in your life to help shape the 'new normal' of our society.

1. Fossil fuels get more subsidies

The fossil fuel industry is global — it has its claws in practically everything. And although there have been some subsidies given to renewables, the ones that coal, oil and gas receive far outweigh them. In fact, fossil fuels get four times the subsidies of renewables. Politics isn't everyone's favourite subject, but it's important for kids to understand how political policies shape our world. This is one place where you can start.

2. Water wheels were once used to generate power

Water is one of the oldest power generators. You can work with your kids to help them build one at home. This is a straightforward but efficient way to explain how water can create power, even if the wheels are not the most conventional method. They're so simple that you can easily make them with things from around the house. Even preschoolers would love to be involved.

3. A dam is a great visual for the power grid

Water is one of those never-ending renewable energy sources that we often forget about. If you want to teach kids about the electricity grid, visit a dam. It's a huge visual representation of a battery that illustrates the energy the world needs. The lake formed by the dam is, essentially, the battery storage. When energy demand is high, water tumbles through the dam into the small river below. When demand is low, water is pumped back up into the dam, creating more energy reserves. Take a tour if you can.

4. Solar and wind industries are creating jobs

Clearly beneficial to both the economy and the Earth, beneficial facts about renewable energy point to the wind and sun. Green energy is a new and upcoming industry. It has already created jobs and will continue as demand grows, and the power grids adapt. Now, solar and wind industries are creating jobs much faster than the rest of the economy. More jobs mean more work and more opportunities. The industry now employs four million people and is only expected to continue growing.

5. Costa Rica went entirely renewable

Well, almost. The residents still drive gas-powered cars, but the electrical grid used 98% renewables for the year. That's an incredible feat, and one that larger countries have been unable to accomplish. Of course, the smaller population and the pleasant temperatures have some effect but does little to dampen the accomplishment. Costa Rica has become a positive example to encourage larger nations to follow suit.

6. One wind source could power a small town

A single wind turbine, if properly placed and utilized, could power 1,400 houses. Wind farms are becoming more popular, especially when they're put in the ocean. The constant breezes on the water create a nearly constant supply of energy, but it's important to place them in an area that isn't too susceptible to severe storms.

7. Green energy is more reliable than a power grid

The electricity grid fluctuates by the hour and season. This is one of the facts about renewable energy that many people don't know. Predicting the needs of people is a huge part of electricians' jobs, but it's not a perfect system. Storms often create power outages, sometimes for weeks. Green energy doesn't supply power from a single source. It's often spread out and uses multiple power generation methods.

8. The Sun is all we need

The best renewable energy source is the Sun. If we invested in solar power and maximized its use, we could power the entire world from the Sun. The best part is, you wouldn't have to lay all those solar panels in anyone's backyard. A large swath of desert could do it, but you could take it even farther. The Moon is a perfect place to harvest solar energy from. With robots and rovers, it's entirely possible to lay down acre after acre of solar panels. What kid wouldn't get excited about that?

Non-renewable energy facts

The non-renewable energy facts are the direct opposites of the facts and benefits of renewable energy mentioned above. Think about them from a different perspective and you can easily determine why these facts about renewable energy are important.

The best fact about renewable energy is in its name – it's renewable. Introduce your kids to the whole idea of how it works, let them experiment with it and get them excited about the future. That's the surest way to protect the industry and the planet.

4 October 2018

www.greenchildmagazine.com

Scottish wind power smashes 100% production threshold

Scottish wind power produced more than 100% of the threshold for the first time, generating enough energy to power six million homes.

By David McPhee

The National Grid energy requirement for November saw both on and offshore wind produce more than the required demand on 20 of 30 days.

Powering 109% of the total energy requirement, the new figures set a new record for wind generation in Scotland.

Gina Hanrahan, head of policy at WWF Scotland said: 'Wind power breaking through the magic 100% threshold is truly momentous.

'For months output has flirted around the 97% mark, so it's fantastic to reach this milestone. It's also worth noting that 20 out of 30 days, wind production outstripped demand.

'Most of this is onshore wind, which we know is popular, cheap and effective. But the UK government needs to allow it to compete with other technologies, by unlocking market access for onshore wind if it's to realise its full potential.'

The National Grid said the best day for wind generation was November 28 which produced 116,599 megawatts (MW) – enough to power 9.59 million homes.

Adam Forsyth, alternative energy and resource research efficiency analyst at Cantor Fitzgerald, last night said the signs were 'encouraging in Scotland', but pointed out that storage technology is not keeping pace with generation.

He said: 'We're definitely about to see an increase in the renewables energy mix. If we have better available storage then it doesn't matter about when the wind blows. On the whole we need to have more renewables and less intermittency moving forward.

Mr Forsyth added that he also expects tidal energy to become a 'greater part of the energy mix in Scotland'.

He said the choice by firms like Simec Atlantis Energy to update and invest in new technology will allow them to become a bigger part of the energy story in Scotland.

He added: 'Having a more mixed energy generation goes a long way to solving issues around intermittency.'

10 December 2018

How we can use wind power when there's no wind

By Mal Chadwick

Anyone who's a fan of clean energy has heard the argument that 'the wind doesn't always blow and the sun doesn't always shine'. This is called 'intermittency' – the idea that you might not get a constant flow of power from a particular source.

All energy sources are intermittent (coal plants can trip, pipelines can explode, and nuclear reactors can be invaded by jellyfish), but it's a particular challenge for renewable energy.

The wind doesn't always blow – and that's ok

Some people think this means we'll always need lots of giant coal, gas or nuclear power stations (sometimes called 'baseload') to keep the lights on. But that's not really how it works anymore.

Luckily, internet commenters weren't the first to discover weather or night-time. In fact, people have been working on solving intermittency for a while. Thanks to them, we now know that **a good mix of renewable energy can do the job**. Here's how…

1. Wire countries together to share power

Imagine if we could pinch a bit of solar from Spain when we're low on power. Well, imagine no more.

We've already built a few undersea cables called 'interconnectors' across to mainland Europe. These allow us to **share energy supplies with other countries**, and there are plenty more on the way. So if the wind drops in the UK, we can ask our friends in Denmark to share their energy with us.

2. Use giant batteries to store power

If we can store energy on a large scale, **we don't need the wind to be blowing all the time**. And this is already happening – massive batteries are popping up all around the country.

You can even let the grid use your electric car battery while it's plugged in, helping to balance the scales in exchange for free charging.

But batteries aren't the only way to store power – there are all sorts of other systems in the works, from compressed air to molten salt, to giant flywheels spinning in a vacuum. We don't know for sure which of these technologies will take off, but because the government has already decided that energy storage is a big deal, it'll soon be as normal as plugging your phone in when you go to bed.

3. Shift power demand away from peak times

So far we've talked about sharing and storing the energy we produce, but **what about the energy we're using?**

In the UK, demand for electricity peaks during cold winter evenings. These peaks put a lot of strain on the system. But now we've got the technology to control the other side of the equation, shifting some of that demand to times when there's more spare power. And **the more demand we can shift, the fewer giant power stations we need**.

We've now got the technology to shift demand to times when there's more spare power

For example, supermarkets could automatically turn their fridges down during those peak times, helping to balance the grid. And because fridges can hold their temperature for a while, the food stays fresh.

Again, **this isn't theoretical** – it's called 'demand side response', and it's already happening in a few places. Sainsbury's are using their fridges to help balance the grid, and Marriott Hotels have trialled a similar setup with their air conditioners. Heavy industry is getting in on the action too: road materials company Aggregate Industries has hooked up their giant bitumen tanks to an AI-based demand response system.

The last resort: flexible gas plants

As we get better at all this, the grid should stay in good shape – even as renewable energy keeps growing. But we still need to be ready for a **worst-case scenario** where the scales just won't balance.

And for those moments we'll keep a little bit of flexible gas power in reserve to make up the difference. And **this doesn't have to mean burning fossil fuels**. We can even use a technology called 'power to gas' to make this process renewable powered too.

This doesn't have to mean burning fossil fuels. A technology called 'power to gas' can make this process renewable powered.

But either way, once we get those other bits right, this will hardly ever be needed. **Renewable energy is already powering our lives.** And now we can balance supply and demand, there's no reason why it can't keep the lights on.

12 January 2018

About 15% of the UK's electricity comes from wind

Claim: A quarter of the energy we use in Britain today comes from wind.

Conclusion: That's a snapshot picture when wind is generating an unusually high amount of electricity. Over the course of a year, about 15% of the UK's electricity comes from wind power.

'And a quarter of the energy we are using in Britain today has come from wind.'

Owen Smith MP, 1 March 2018

At the time of writing, wind power is currently generating about a quarter of the UK's electricity, about the same as gas. That's according to live data taken from the National Grid and electricity administrators Elexon. That includes both on- and offshore wind generation.

But that's unusually high, because of the recent weather. The amount of electricity generated from wind is volatile.

Wind generated something like 15% of the UK's electricity generation in 2017, and 13% using the latest official figures for the 12 months up to September last year. Gas remains the single biggest source of electricity by far.

The same graph five years ago looked very different to now. The amount of electricity we get from wind and other renewables has been increasing, while the amount we get from coal has plummeted.

That's because of a mix of increased wind capacity, falling coal production and market pressures on coal.

The large drop in coal is because of closures to coal plants in the UK, the conversion of existing coal generators to biomass, and market factors. As the Department for Business, Energy & Industrial Strategy says: 'Whilst fuel costs for coal-fired generation are lower than for gas, emissions from coal are higher so generators must pay a greater carbon price per GWh produced.'

This has been combined with a large increase in the UK's wind generation capacity. The trade association RenewableUK provided us with figures showing the total installed capacity of wind generators has doubled since 2012.

Some of the electricity supplied in the UK comes from imports from the continent. The UK currently imports far more electricity than it exports.

2 March 2018

www.fullfact.org

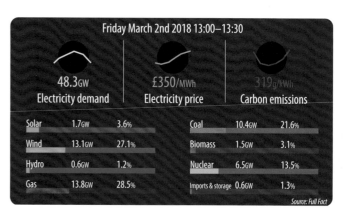

Friday March 2nd 2018 13:00–13:30				
48.3GW Electricity demand	£350/MWh Electricity price	319g/kWh Carbon emissions		
Solar	1.7GW	3.6%	Coal	10.4GW	21.6%
Wind	13.1GW	27.1%	Biomass	1.5GW	3.1%
Hydro	0.6GW	1.2%	Nuclear	6.5GW	13.5%
Gas	13.8GW	28.5%	Imports & storage	0.6GW	1.3%

Source: Full Fact

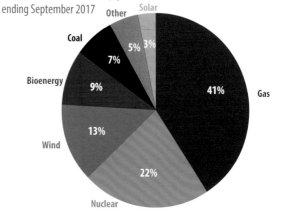

UK electricity generation

Proportion of total electricity generated from different sources in the 12 months ending September 2017

Gas 41%, Nuclear 22%, Wind 13%, Bioenergy 9%, Coal 7%, Other 5%, Solar 3%

Source: Department for Business, Energy & Industrial Strategy, Energy trends: electricity, tables 5.1 and 6.1

The wind beneath our wings

Change in proportion of electricity generated from different sources, comparing the 12 months ending September 2011 to the same in 2017

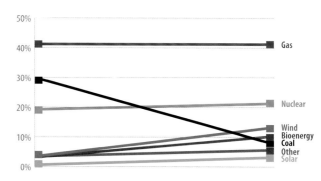

Source: Department for Business, Energy & Industrial Strategy, Energy trends: electricity, tables 5.1 and 6.1

The way the wind blows

Newly installed UK wind capacity in Gigawatts per year, as at year end

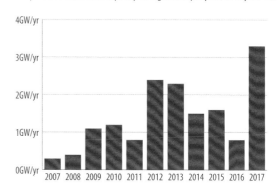

Source: Figures provided to Full Fact by Renewables UK (March 2018)

Ocean energy: Time to unleash the current

Gesine Meissner MEP, Special Envoy to the President of the European Parliament on Maritime Policy, shares her thoughts on ocean energy and why now is the time to unleash the current so that renewable technologies in Europe are pushed forward.

Unlike wind and solar power, ocean energy is reliable and predictable. Waves are available all year round, the tides are fixed and the salt content of the sea is almost constant. The amount of CO2 released during production is low and the associated electricity costs are as well. Studies show that so-called Blue Energy has the potential of meeting 10% of the EU's power demand by 2050. This would cover the daily electricity needs of 76 million households. Almost all major technology developers have their headquarters in Europe. The sector should be worth €53 billion annually in 2050. So, what is holding back this energy source from flourishing?

While European companies are a clear frontrunner in this technology and account for a large share of patents globally, it remains a comparably expensive technology. Most ocean energy projects are still in their test phase and initial funding is difficult to obtain. Beyond this obstacle, there is also a lack of appropriate infrastructure. In fact, transmission networks to distribute the energy produced have not been sufficiently expanded yet. Furthermore, the impact of certain projects on the surrounding ecosystems (animals, algae and coastal structures, etc.) are still partly unknown and are becoming an increasing point of concern when planning. There is a similar concern regarding the overall lifecycle carbon cost of this energy.

I would, however, argue that the real obstacle to success is its lack of visibility at the political level. And this, despite Europe being a leader in this field. But while scientists and policy experts have recognised the potential of ocean energy a long time ago, it is still not acknowledged by the mainstream policymakers.

As witnessed with the development of other renewable technologies in Europe, initial public investment is key to attract and accelerate private investment. While the projects that currently exist are mainly local or national ones, developing a real European interest could provide the push needed for the sustainable commercialisation of

ocean energy. Many European countries have ocean access and as such, they have a shared interest in pushing for the development of this type of energy.

At the European level, things have already been moving in the right direction. Since 2014, the Commission has invested €150 million in ocean energy through its research and innovation framework programme, Horizon 2020. But we need further investment in researching, developing and testing blue energy. This is why I am actively pushing for ocean research to remain a central element in the new research programme, Horizon Europe. The legislative package 'Clean Energy Package for all Europeans', sets a renewable energy target for the EU by 2030 of 32%. This is also a chance for ocean energy to contribute to achieving our climate goals, especially as ocean energy can work as a grid stabiliser in combination with other renewable kinds of energy.

Moreover, we need to ensure that our oceans are not further damaged by marine pollution. While waves and tides are more reliable than other renewable energy sources, the ocean is also impacted by climate change. A recent study by the Geomar Helmholtz Center for Ocean Research shows that the warmer the ocean, the weaker the ocean circulation. This is another area where the European policymakers can play a key role.

Once the right political framework is in place, Blue Energy can become a profitable and sustainable economic sector. But this also requires a stronger involvement of the private sector, as well as the willingness of private investors to take risks. It is our job, as European leaders, to create the confidence needed for such investment.

29 October 2018

Renewable energy: Report claims large hydropower projects damaging to the environment

The report's author said: 'Large hydropower doesn't have a future, that is our blunt conclusion.'

By Albert Evans

A new study has concluded large-scale hydropower projects in European and North America have been disastrous for the environment and that similar projects in the developing world should not be built.

Hydropower makes up nearly 71 per cent of the world's renewable energy and has long been viewed as one of the most dependable forms of green electricity. But in Europe and North America, many dams are being removed either for safety or environmental reasons.

The scientists' report sets out a litany of problems with many of the dams built in the last century, many of which were critical to the economic development of their respective countries.

The vast majority were more expensive than they planned, destroyed the local environment and contributed to climate change by releasing greenhouse gases from the land they flood.

One of the report's authors Professor Emilio Moran from Michigan State University told BBC News.'They make a rosy picture of the benefits, which are not fulfilled and the costs are ignored and passed on to society much later.'

While developing countries are reducing the amount of hydroelectricity they use, vast projects are in the pipelines in much of the developing world.

The authors say that with huge pressure on countries to press ahead with renewable energy developments, a mix of energy sources including hydro is the most sustainable approach.

The report said that frequently the benefits of the power produced does not feed through to local residents, who are left to face the environmental problems and food shortages they can cause.

Professor Moran said: 'Large hydropower doesn't have a future, that is our blunt conclusion. To keep hydropower as part of the mix in the 21st century we should combine multiple sources of renewable energy'.

'There should be more investment in solar, wind and biomass, and hydro when appropriate – as long as we hold them to rigorous standards where the costs and benefits are truly transparent.'

6 November 2018

Majority of UK public want to install solar panels, poll finds

More than 70% would make homes more energy efficient given government support

By Adam Vaughan

More than half of the British public would install solar panels and home batteries to tackle climate change if there was greater assistance from the government, polling has found.

While many have already made their home more energy efficient, 62% said they wanted to fit solar and a surprisingly high 60% would buy an energy storage device such as those sold by Tesla.

An even greater number – 71% – would join a local energy scheme such as a community windfarm or solar panel collective, according to the YouGov survey.

The results run counter to the government's approach to climate change and energy, which favours large-scale power generation such as nuclear plants and offshore windfarms.

Community energy projects have flatlined in the face of government subsidy cuts and tax changes, while incentives for household solar will expire next year without a replacement. There is no support for people considering a home battery.

James Thornton, CEO of environmental law group ClientEarth, which commissioned the research, said: 'Government policy is plainly at odds with public sentiment – and its own ambition to tackle climate change – as far as our energy sources are concerned.

'People want to know more and take ownership of how they get their energy – that's clearly demonstrated by the broad support in the poll for household solar and community energy schemes.'

Solar installers have told *The Guardian* that, increasingly, people are also opting for home batteries when they buy solar.

Energy storage is also making inroads at utility scale. On Monday, water company Anglian Water will announce it has bought a 300 kilowatt hour (kWh) storage system from UK-based firm redT, for use at a water treatment site alongside solar panels.

Solar was ranked the most popular of all energy sources in the ClientEarth survey, while gas was second worst behind coal.

More than two-thirds (68%) thought the big six energy suppliers' market dominance should be broken up to allow smaller clean energy firms to grow. Exactly half said they would move their pensions to avoid fossil fuel investments, a figure that rose to 59% for 18–34-year-olds.

20 August 2018

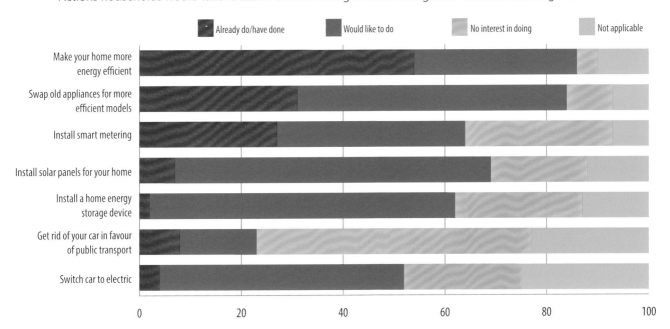

A majority of Britons want to install solar and home batteries

Actions households would take to tackle climate change if there was greater assistance from government

Legend: Already do/have done — Would like to do — No interest in doing — Not applicable

- Make your home more energy efficient
- Swap old appliances for more efficient models
- Install smart metering
- Install solar panels for your home
- Install a home energy storage device
- Get rid of your car in favour of public transport
- Switch car to electric

Source: ClientEarth/YouGov

Is the UK sunny enough for solar panels to work?

Despite its grey weather, the UK is a good place to install solar panels.

Since solar panels rely on the energy of the Sun to produce electricity, they won't work in the cloudy and grey UK. **WRONG**. Picture one of those solar calculators you might have someday used at school. Did they stop working on cloudy days? No, they didn't.

The technology used to build a solar panel is the same as that used for a solar calculator, just a bit more complex. So, contrary to what many would think, despite its reputation for having grey and cloudy weather, the UK has more than enough sunlight to power solar panels. It actually gets the same amount of solar radiation as certain areas in France or Spain, which are meant to have more Mediterranean climate. In fact, the UK gets around 60% of the solar radiation found in the Equator and, in some areas in the south, it receives a comparable amount of sunlight to that in Germany, one of the biggest markets for photovoltaics in the world.

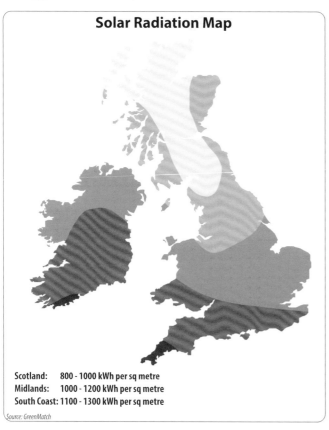

Solar Radiation Map

Scotland: 800 - 1000 kWh per sq metre
Midlands: 1000 - 1200 kWh per sq metre
South Coast: 1100 - 1300 kWh per sq metre

Source: GreenMatch

So, the good news is that you don't necessarily need the Sun hitting hard for solar panels to work. Even though they obviously produce more power during a sunny day, they can still produce a considerable amount of energy when the days are cloudy. Solar PVs use light to produce electricity, not heat.

What about wind? More good news

Although high wind speeds commonly found in the UK could be a concern to anyone wondering about the efficiency of installing solar panels in this area, wind can actually be beneficial for domestic solar panels since it can serve as a cooling mechanism for the PV modules, leading to an increase in their efficiency.

And...what about snow?

Due to the greenhouse effect and other crazy things on earth which humans have lost control of, weather has become more extreme in every corner of the planet. And yes, the United Kingdom is not an exception. Winters in the UK have turned a little more aggressive, bringing heavy snow to large areas of the country every year. This can be a problem for solar panels since it might accelerate their ageing.

But you shouldn't panic. There's always a solution. Solar panels should be installed at an angle that maximises their performance. What is the ideal angle? Don't worry if you don't have the answer, there are experts ready to help you with this.

Going back to snow and solar panels, a bit of snow will easily melt with residue heat from the panel. In this case, you shouldn't worry at all. If there is more than just a bit of snow, and it doesn't melt fast, it will unfortunately prevent the panel from working to its full potential. In this case, some hand work is required. A little brushing of the panel as fast as possible is all it takes though. Moreover, if you live in an area where heavy snow is a frequent scenario, there are some kind of treatments that make the panel's surface slippier making it harder for the snow to accumulate.

The good news is that some snow lying around can in fact increase the panel's performance by reflecting the Sun's photons back up from the ground.

So am I crazy if I want to install solar panels in the UK?

Not at all! In fact, there were around 230,000 solar power projects in the UK by the end of 2011. The reduction in the cost of photovoltaic and the Feed-in Tariff (FIT) introduced in 2010 helped to make this trend more popular. Despite the fact the FIT rate has been recently reduced, you can still make a very reasonable and cost-efficient investment.

So no excuses accepted. Wind, snow, grey sky, nothing of this will beat the solar panels. You should request customized assessment to get the ideal solution for your property. There are different options to choose from and also better ways than others to install the desired equipment.

9 March 2018

Homeowners trapped by 25-year solar panel contracts

Householders who lease their roofs to power firms find it hard or costly to move home.

By Anna Tims

Julie Griffiths* wanted to reduce her carbon footprint by installing solar panels. The cost would have been a prohibitive £12,000, so she signed a deal to lease part of her roof to a solar power company, which would fit the panels for free. It would pocket the newly introduced Feed-in Tariffs (FITs) – subsidies paid by the government for the electricity generated. She, meanwhile, would have lower energy bills. At the end of the 25 years, the panels and the tariffs would be hers.

It seemed a win-win situation until recently, when she needed to sell the house. Her buyer's mortgage application was refused because of the lease agreement, which had effectively signed over a large part of the roof to the solar company.

'A clause in the lease allowed us to buy out the panels for a fee to compensate the company for the loss of their FITs,' she says. 'We were prepared to do it to be able to sell our house and move on with our lives, but the company had passed the management of the panels on to an agent, who seemed reluctant to let us proceed.'

Eventually, after *The Observer* intervened, Griffiths was allowed to buy the panels for £20,500, an uncosted sum that she was told was non-negotiable. Not only was this nearly double the price she'd have paid to install the system herself, but she had also missed out on nearly eight years of FITs worth, around £7,300. All in all, the 'free' system has left her around £16,000 out of pocket.

The government introduced the generous incentives in 2010. The FITs, funded by a levy on all energy bills, have encouraged 800,000 households to go solar, but they have also spawned a multitude of startups that have exploited homeowners.

Homeowners who wanted to do their bit for the environment but could not afford the outlay were promised up to 50% off their bills if they signed over the airspace above their roof for 25 years. For the startups it was a bonanza. Payouts would earn them an average of £23,000, more than triple their investment. But unscrupulous contracts obliged owners to seek permission if they wanted to extend or sell their home, or compensate them if the panels were temporarily removed for roof repairs.

Since 2012, as installation costs have plummeted, the feed-in rates have been slashed for new installations by 90%, and they will be abolished for those who install solar panels after 31 March next year. The profiteering startups have all but disappeared, but their legacy will blight the lives of homeowners and unwitting buyers for two more decades.

Many are discovering the high price of their 'free' deal as they try to sell. The 25-year leases apply to the property regardless of who owns it, and they have to find a buyer willing to take on the remaining years. And even if a buyer is happy, mortgage lenders may not be. The deal is treated as a leasehold, and contracts skewed in favour of the company are deemed risky by banks and building societies.

Claire Hunt discovered that her elderly father had signed up to a leased solar installation when his roof began leaking. He had been talked into the 25-year contract in 2011, when FITs were at their peak, and the company has since ceased trading.

'We were directed to another company which has apparently taken over the contract, but we've been unable to engage with them by phone or in writing,' says Hunt. 'It is only because of the leak that I have finally had sight of the contract and I am devastated to find there isn't even a buyout clause to end it. We are therefore arranging to remove the solar panels ourselves and will put them back when completed. A very expensive fix.'

Fiona Baker is similarly trapped after agreeing to leased solar panels in 2010. 'I am retired with MS and myotonic dystrophy, and need to sell my house, but the buyer won't go ahead unless I have the panels removed,' she says. 'This is apparently not possible until the lease expires at the end of the 25 years.'

Meanwhile, Simon Norris is unable to remortgage because the firm that installed his leased panels appears to have breached building regulations. 'The lender wanted a structural survey, which concluded that the roof should have been reinforced before the panels were fitted,' he says.

Most lenders will agree a loan on a property with leased solar panels provided the contract meets certain conditions, one of them being that the installing company be accredited, the installation be approved and insured, and panels removable without penalties for missed FIT payments. Crucially, a lender's permission needs to be obtained if an existing borrower decides to install leased panels.

In the race to profit from FITs, many companies ignored these conditions. They also did not inform customers of the full value of subsidies they would be entitled to if they bought the panels outright or the implications for current or future mortgages as required by the Renewable Energy Consumer Code (RECC).

Since most of the original companies have ceased trading, it can be difficult for sellers and buyers to find out who owns their panels. Jim Cowan recently bought his house without receiving the contract for the leased panels. He can't find out who owns them since the installing company no longer exists and may find himself unable to remortgage or sell. Householders in this predicament have to submit a subject access request to energy regulator Ofgem's FIT register team and provide proof that they are the property owners.

Now that FITs have all but dried up, profiteering has taken a new form. This time companies are targeting those who bought panels outright between 2010 and 2015 and continue to receive the old higher-rate subsidies. They offer a tempting-looking lump sum in return for the FIT for the remainder of the 25 years. However, the offered price is invariably less, sometimes by 75%, than the sum customers would receive from their FIT and the conditions are often similar to the lease stitch-up.

Government subsidies on solar panels will cease at the end of next March. Miss the deadline and it would take a working household up to 70 years instead of the current 20 to recoup the average £6,500 cost. Those hoping to install panels should get quotes only from companies listed on the RECC website and request a written estimate of the output from the system and the financial benefits. Anyone buying a house with leased solar panels should ensure the contract complies with the minimum requirements laid down by the Council of Mortgage Lenders.

* All names have been changed

25 November 2018

Why is it so hard to get people to switch to renewable energy?

Renewable energy at home – such as solar panels on the roof – can help save energy costs but also reduce a little our impact on the environment in terms of climate change. With such a win-win solution, why are we not all making the switch, asks Emily Folk.

Renewable energy sources are becoming increasingly more easily available every year. The clean energy revolution could not be timelier – with the effects of global climate change and our planet's resource scarcity coming into full view.

Despite the confluence of these forces – green energy technology and a heightened public perception of climate change – the movement toward personal renewable energy sources has been sluggish. People are still wary of making the switch. Why is this?

There are a few primary obstacles consumers face when making the green transition. This article will explore a few of the most challenging ones.

Electric company

The assumption for many would-be green energy converts is that the initial price ceiling is high. If you have money saved, then investing in green energy and negating an electric bill would be an obvious next step.

But misinformation and shifting prices leave many people confused on exactly how much renewable energy costs, much less how it works. Many consumers fail to investigate properly and assume that installation and maintenance costs of green energy are far outside their budget.

In reality, the prices of renewable energy are highly variable but rarely backbreaking. The basic package for personal solar panels, for instance, typically runs between $9,000 and $12,000 after tax credits, and covers both panels and installation. Though the return on investment may take years, the package cost is much less than many imagine.

Studies have shown time and again that people remain concerned about the environment in the abstract but are not willing to personally modify their lives.

In the case of switching to renewable energy, the perceived shift is a large one: the electric company has to be called, the installation will take time, and then there's the aforementioned investment.

Solar gear

The ordinary fossil fuel system is streamlined and accessible enough that people don't want the inconvenience and uncertainty of switching.

Today, however, the switch is significantly easier than most people think. If you want to save money by self-installing, it may take a bit longer.

Lucky for you, there are plenty of online resources which can smooth the entire process. Most converts opt for the assisted installation packages which cost more, but they also cut out the hassle of drilling through your roof.

Solar is a highly modular system, meaning it can be customised or improved on the go. Starting with a small package of panels can be a good start for those unsure of their options, and can be added to as the prices of solar gear drop.

Likewise, changes to the roof of your house can be made without damage or hassle to the system in place. Changing from classic shingles to a metal roof allow for a more durable panel anchor, and can cut summer cooling costs by up to 20 percent as well.

Biggest culprit

Of course, this only refers to solar energy, and there are plenty of other clean energy solutions with varying degrees of difficulty for installation. Geothermal – better known as 'heat pumps' for individual use – are nearly impossible to DIY, but also have a developed and cost-effective system for assisted installation.

Wind turbines, like solar panels, depend on the size and ambition of the project. However, wind farms have begun sourcing their energy to far-off cities and individual residences, meaning very little has to be done on the part of the individual. Prices here are typically competitive with standard electricity costs.

Many of these conversion barriers are perception-based, which is no surprise. There is an overwhelming amount of misinformation and misconception regarding renewable energy.

While the two obstacles above are both rooted in misconception, the biggest culprit is that of reliability, a stigma that has stuck with clean energy since its popularisation.

You have solar panels strapped to your roof – so what happens when it rains? Obviously, during that period, there is no solar energy for the panels to absorb, and the meter will continue to run up.

Daunting task

However, even during periods of overcast skies, panels can operate – though tougher when not at full capacity – and the energy generated during the sunny days is enough to offset the cost.

As for wind, the power generated is often even more reliable than solar, and sourced energy will remain as steady a cost as standard electricity.

However, the reliability fear cuts deeper than this. Stories continue to circulate from the early days of solar energy, including issues with electrical grids powered by clean energy.

Today, with the continued research and advancement in the field of renewable energy, grids powered in part or fully by renewable sources are typically as reliable as those using standard power generation. In fact, a 2017 Department of Energy in the US report confirmed clean energy as a reliable and safe source of power for American homes and businesses.

Making the switch to clean energy can be a daunting task, and there will inevitably be some lifestyle modification and monetary investment involved.

However, given the amount of misinformation floating around, anybody who is even marginally interested in switching to green energy should complete some research on the topic. You might be surprised at how easy and accessible the process truly is.

13 March 2018

Do renewable energy technologies need government subsidies?

To what extent are different energy sources subsidised around the world?

By Georgina Kyriacou and Samuela Bassi

Renewable energy comes from sources that are not depleted when used but are replenished naturally. They include wind, solar, hydro, tidal, wave and geothermal energy, and are generally used for power generation or heat production. Their use has grown rapidly in recent years, largely to take the place of fossil fuels as countries try to reduce their carbon dioxide emissions in the fight against climate change.

The International Energy Agency (IEA) has calculated that subsidies to aid the deployment of renewable energy technologies amounted to US$140 billion in 2016. Countries within the Organisation for Economic Co-operation and Development (OECD) subsidise green energy more than poorer, non-OECD countries.

While renewables are often criticised for being heavily subsidised, in fact fossil fuels and nuclear power receive more financial support. The IEA calculated that fossil fuels received about US$260 billion in 2016 and the United Nations Framework Convention on Climate Change (UNFCCC) has criticised these subsidies for hampering progress on reducing emissions. Some describe fossil fuels as receiving preferential treatment politically. However, many countries are committing to phasing out fossil fuel subsidies and their amount decreased by 15% between 2015 and 2016.

There is also an 'implicit' subsidy to fossil fuels, because the price of power generated by fossil fuels does not reflect the environmental costs they generate in the form of climate change and local air pollution.

What role can subsidies play?

The use of subsidies is motivated by the need to address market failures, such as to address the price disparity with fossil fuels when environmental costs are not accounted for. Other market failures affecting renewable energy sources, such as spill-overs from research and development and economies of scale, may also warrant a higher price in early years to induce more innovation.

Subsidies to renewables have been credited with increasing innovation, lowering costs and expanding the energy mix – roles also played by early subsidies to fossil fuels, which were greater than those made to renewables at the same stage of development.

By increasing the deployment of renewables, subsidies have played a role in reducing reliance on fossil fuels. This is very important for reducing greenhouse gas emissions and restricting global temperature rise. International Renewable Energy Agency (IRENA) analysis in 2017 showed that renewable energy (with energy efficiency) could meet 90% of the Paris Agreement's energy-related goals, but that to do so further technological breakthroughs and new business models will be required.

Benefits can also be measured in other ways. Researchers from Berkeley National Laboratory found that using wind and solar energy in place of fossil fuels helped avoid between 3,000 and 12,700 premature deaths in the US between 2007 and 2015, saving US$35–220 billion. The researchers concluded that the monetary value from improved air quality and climate benefits were about equal to or more than the cost of government subsidies to wind and solar.

Are subsidies still needed?

Today renewable sources of electricity are becoming cost-competitive with fossil fuels and nuclear power and will soon no longer need subsidies. In the context of the European Union, for example, analysis has suggested that countries should focus on carbon pricing rather than subsidies for low-carbon electricity to achieve further reductions in power sector emissions, and as a more cost-effective measure. This is backed by other research which

has found that in a time of more stringent climate policy, strong carbon pricing is the preferable policy instrument to encourage demand for renewable energy, but that to meet less ambitious climate targets, using subsidies alone would be successful in stimulating further development of clean technologies.

In the UK, subsidies have led to a significant increase in the deployment of renewables. This in turn has led to a rapid decrease of the cost of some of these technologies. Notably, offshore wind projects commissioned in 2022/23 will sell their electricity at £57.50/megawatt hour (MWh). This is cheaper than the average cost of generating electricity from gas, and well below the price of long-term contracts for new nuclear power, which is £92.50/MWh. Recent analysis backed the UK government's view that further subsidies should be time-bound and removed once the relevant obstacles and market failures have been overcome. This will include phasing out fossil fuel subsidies, as recommended by the International Monetary Fund (IMF).

14 May 2018

www.lse.ac.uk/GranthamInstitute

EU countries by production of clean energy

Top ten EU countries by production of clean energy

n 2014, the amount of renewable energy share reached 14% of total energy consumption and was estimated to reach 16.4% in 2015. The goal of 20% for the European Union is now within a reachable distance. But, it is now time for all member states to keep up their efforts to reach the goal.

All in all, Europe is performing admirably in its organisation of renewables. In 2011, renewables created 21.7% of the EU's power; after three years, this figure has achieved 27.5% and is expected to grow to 50% by 2030. The EU's underlying endeavours in advancing the utilisation of renewables encouraged this. Proceeded by the development which brought down sustainable costs: the costs of photovoltaic modules fell by 80% between the end of 2009 and late 2015. Renewables have now moved toward becoming cost-focused, and even sometimes significantly less expensive than fossil fuels.

The turnover in the renewable energy sector claimed an amount of €144 billion in 2014 and over one million jobs were created, thus it plays an important part in the European economy. European speculations have dropped by a large part since 2011 to €44 billion a year ago, while worldwide interests in sustainable power source have reached above €260 billion.

Why are renewables a key segment?

The Renewable Energy Directive has been and will keep on being, a focal component of the energy union strategy and a key driver for clean energy. With the point of making the EU the world's number one spot for renewables. Renewables have played a noteworthy part in energy security. The predicted contribution to savings by reducing the number of imported non-renewables was €16 billion in 2015 and a

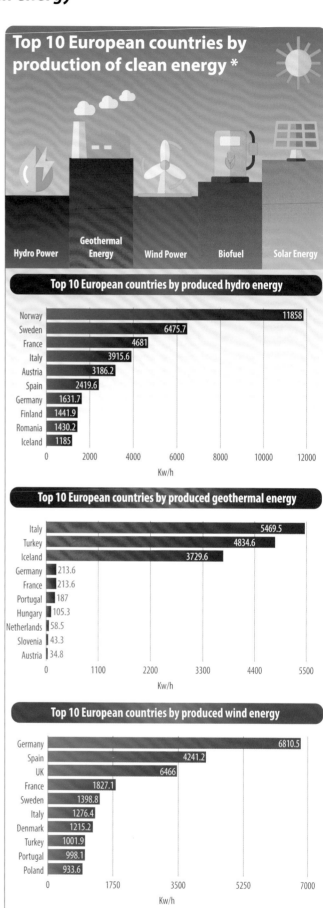

Top 10 European countries by production of clean energy *

Hydro Power | Geothermal Energy | Wind Power | Biofuel | Solar Energy

Top 10 European countries by produced hydro energy

Country	Value
Norway	11858
Sweden	6475.7
France	4681
Italy	3915.6
Austria	3186.2
Spain	2419.6
Germany	1631.7
Finland	1441.9
Romania	1430.2
Iceland	1185

Kw/h (0, 2000, 4000, 6000, 8000, 10000, 12000)

Top 10 European countries by produced geothermal energy

Country	Value
Italy	5469.5
Turkey	4834.6
Iceland	3729.6
Germany	213.6
France	213.6
Portugal	187
Hungary	105.3
Netherlands	58.5
Slovenia	43.3
Austria	34.8

Kw/h (0, 1100, 2200, 3300, 4400, 5500)

Top 10 European countries by produced wind energy

Country	Value
Germany	6810.5
Spain	4241.2
UK	6466
France	1827.1
Sweden	1398.8
Italy	1276.4
Denmark	1215.2
Turkey	1001.9
Portugal	998.1
Poland	933.6

Kw/h (0, 1750, 3500, 5250, 7000)

* 1000 tonnes of oil equivalent   Sources: http://ec.europa.eu/eurosta, https://windeurope.org, & Shambo Nagpall

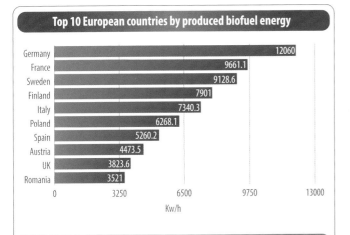

Top 10 European countries by produced biofuel energy

Country	Kw/h
Germany	12060
France	9661.1
Sweden	9128.6
Finland	7901
Italy	7340.3
Poland	6268.1
Spain	5260.2
Austria	4473.5
UK	3823.6
Romania	3521

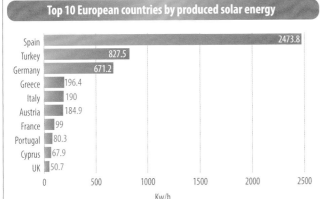

Top 10 European countries by produced solar energy

Country	Kw/h
Spain	2473.8
Turkey	827.5
Germany	671.2
Greece	196.4
Italy	190
Austria	184.9
France	99
Portugal	80.3
Cyprus	67.9
UK	50.7

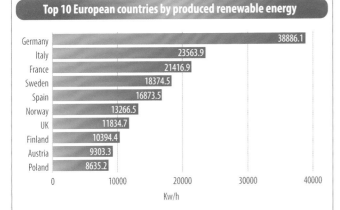

Top 10 European countries by produced renewable energy

Country	Kw/h
Germany	38886.1
Italy	23563.9
France	21416.9
Sweden	18374.5
Spain	16873.5
Norway	13266.5
UK	11834.7
Finland	10394.4
Austria	9303.3
Poland	8635.2

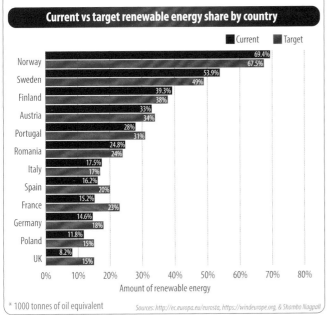

Current vs target renewable energy share by country

Current / Target

Country	Current	Target
Norway	69.4%	67.5%
Sweden	53.9%	49%
Finland	39.3%	38%
Austria	33%	34%
Portugal	28%	31%
Romania	24.8%	24%
Italy	17.5%	17%
Spain	16.2%	20%
France	15.2%	23%
Germany	14.6%	18%
Poland	11.8%	15%
UK	8.2%	15%

Amount of renewable energy

* 1000 tonnes of oil equivalent

Sources: http://ec.europa.eu/eurosta, https://windeurope.org, & Shambo Nagpall

projected €58 billion in 2030. Because of quick diminishing costs, renewables can be continuously coordinated into the market.

The recast of the renewables mandate together with the market outline proposals will additionally empower the investment of renewables on an equal balance with other energy sources. Renewables run as an inseparable unit with energy proficiency. In the electricity segment, fuel changing from flammable petroleum derivatives to non-burnable renewables could decrease essential energy consumption. In the building part, renewable arrangements can enhance the energy execution of structures in a practical manner. Renewables are one of the backbones of lowering CO_2 emissions.

What is the role of renewables in achieving the Paris climate objectives?

In 2015, renewables added to decreasing EU ozone depleting substance (GHG) outflows by 436 MtCO2eq, Italy's emission equivalent. Renewables assume a noteworthy part in making the EU a worldwide pioneer in development. EU countries hold 30% of the patents in renewable energy globally. A real pioneer move of the EU to stay ahead because the EU will keep the priority on research and innovation in the foreseeable future.

At the 2015 Environmental Change Gathering in Paris, Europe conferred itself to adding to constraining the rise of global temperatures to just 1.5°. Renewables, including energy proficiency, are critical to achieving this objective.

6 November 2018

www.greenmatch.co.uk

Has Spain learned its renewable energy lesson?

What a difference a decade makes. In 2007 Spain was intent on becoming of one the world's green energy leaders. Second only to Germany in Europe for installed solar capacity, they had just opened the world's first commercial solar thermal polar plant close to Seville. And on top of that, the government was offering generous subsidies, promising above market rates for green energy producers to help ensure that more people would invest in renewables. And invest they did – there was a huge development of both wind and solar farms.

But all was not well with Spain's renewable energy sector. So let's take a look at exactly what went wrong and see whether the country has taken sensible steps to stop these problems coming up again in future. In short, let's see whether Spain has learned its renewable energy lesson.

Here, Mike James – a writer and Marbella real estate specialist – discusses the impact made by Spain's attempts to go green.

Spiralling costs

The problem was that this subsidy scheme was appallingly badly structured and Spain began to have an insurmountable deficit between the amount utilities companies were paying to green energy providers, and the amount those companies were getting from their customers.

Much of this was due to the fact that the costs were not passed on to customers, so as the cost of supply went up, the prices for the energy remained very low. At its peak in 2012, Spain lost €7.3 billion and has reached debts of €26 billion. But this is far from being the only problem that Spain's failed green revolution created.

It has been suggested 2.2 jobs were lost for every job that the green energy industry created. And each green job that was created is estimated to have cost Spanish taxpayers an eye-watering $770,000 (and only one in ten of those jobs were permanent). It's clear, then, that Spain got its attempt at promoting renewable energy wrong – very badly wrong.

The 'Sun tax'

It doesn't end there. The Spanish government have even been strongly criticised for the policies that they have used to attempt to get the country out of the situation. They have introduced the first ever so-called 'sun tax' – where new solar installations face heavy taxes which makes it almost impossible for them to be economically viable.

Some have suggested that this tax is reasonable, as those using their own solar energy usually rely on the national grid as a backup if they aren't producing enough solar power. However, others have suggested that the government is merely acting to protect traditional power suppliers who don't want to see the solar industry succeed.

Spain has seen its solar sector go badly downhill since. However, there are signs that in 2017 and beyond, renewables in Spain have reasons to feel positive.

Cautiously optimistic

Things are improving. In January 2015 Spain set a new daytime record, when wind energy accounted for 54 per cent of their total energy used. And in November of that year, wind accounted for more than 70 per cent of total energy. But before we get too optimistic about Spain's turnaround, remember that their Iberian neighbour Portugal recently managed to run purely on renewable energy for four days straight, so clearly there is some serious room for improvement.

It's currently in the balance as to whether they will hit the target for them by the European Union: for renewables to make up a total of at least 20.8 per cent of total energy used by 2020. At the end of 2016 the number was at 17.4 per cent, which means that the country is currently just on target to achieve the goal.

A long way to go

But before we get carried away, it's clear that issues remain that the government will need to sort out. In 2015, Spain's government failed to install a single megawatt of new wind capacity for the first time since the 1980s. To meet the EU's 2020 target, Spain will need an additional 6,400 megawatts. This is achievable as the country has managed to do it before, but that was during the green boom years when market conditions were far better suited to that sort of growth. So clearly Spain still has some learning to do if it is going to find a way to produce green energy without putting itself into further debt.

3 February 2017

Chapter 2

The future of energy

Europe's transition to a green energy future

Sweden is one of the leading lights when it comes to energy transition in Europe. The Nordic country gets more than half of all its energy needs from renewables. By 2045, it plans to be carbon free. The Scandinavian state's success, however, is in stark contrast when compared to others. Despite the EU setting transition targets for 2020, some countries continue to lag behind. For instance, Poland still remains deeply reliant on coal to generate its energy. With so many jobs dependent on the fossil fuel, there is an obvious reluctance to change.

Marek Wystyrk is a former Polish miner. He admits that transitioning to cleaner fuels is necessary but still believes coal has a future.

'I think we have to use our wealth of coal. I'm from a coal-mining family. But I know that we have to facilitate change because of pollution and our climate,' he says.

Marek steered his eldest son, Szymon, towards a high school specialising in green energy. For Szymon, the change can't come soon enough.

'The environment is very important to me, because in the place where I live, I don't need to smoke cigarettes. Just by breathing, it's like smoking ten packets of cigarettes a day… but my generation will make a change. We are starting to do that.'

While the majority of Poland's energy still comes from coal, small steps are being made to generate greener energy in the country through solar farm initiatives.

By 2020, 15 per cent of Poland's energy needs should come from renewables.

There is also a push to make coal cleaner. Krzysztof Kapusta is a researcher at the Clean Coal Technology Centre in Mikolow. The lab is funded by the EU.

Share of energy from renewable sources
(in % of gross final energy consumption)

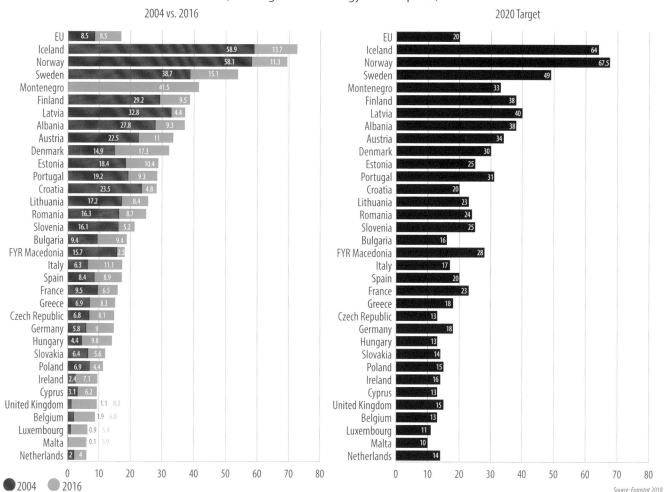

Source: Eurostat 2018

Renewable energy in Europe

 2012
The year when renewables reduced CO2 emissions
by 326 Mt (equivalent to the annual emissions of Spain)

 1 million
The number of people that work in renewables

 €30 billion
The yearly reduction of our fossil fuel imports

2,400
The number of renewable energy
cooperatives in Europe in 2015

 €130 billion
The amount earned by EU renewable companies

80%
The prices that solar panels have fallen in just four years

Source: EuroNews

'Gasification of coal is better than conventional burning because it makes it possible to reduce the environmental impact of the coal utilisation – by removing contaminants such as sulphur and mercury, for example, before coal utilisation.'

At the University of Silesia, which is located in the heart of Poland's coal mining region, Professor Piotr believes change can happen quickly if the government steps in.

'There are a lot of enthusiastic people, they start to use… different kinds of renewable energy. If our government will lead them to be active in this way, the situation will change very fast, I think.'

Europe's energy picture

⇨ Europe has doubled its renewable use in the past 12 years.

⇨ The energy we waste in Europe could power all our buildings' needs.

⇨ Energy efficient products could save families up to 500 Euros a year.

⇨ More than a quarter of the innovations for new tech in renewables are made by European companies.

⇨ The ocean at Europe's doorstep could eventually power 10% of all our demand.

But, the drive towards cleaner energy has not been without its problems. Countries in Europe continue to disagree over the bloc's 2030 transition targets.

Sweden's Energy Minister, Ibrahim Baylan, however, insists change is both inevitable and economically beneficial.

'Coal is not anymore the cheapest way of producing electricity or energy… Solar is! This year, we are seeing off-shore wind being built without any subsidies. So, I think for those countries who are still arguing for fossil fuels… for coal… From economic point of view I cannot understand it anymore.

'Obviously when we made the transition it also created tens of thousands of local jobs… As a politician you have also to see not only the jobs you have today.'

Nevertheless, the gap between EU countries when it comes to generating energy from renewables is significant.

In 2016, Eurostat figures showed that Austria (72.6%) and Sweden (64.9%) produced at least three-fifths of all their electricity from renewable energy sources, while Portugal (54.1%), Denmark (53.7%) and Latvia (51.3%) produced more than half.

At the opposite end of the scale, the lowest proportions of renewables were registered in Luxembourg (5.4%), Malta and the Netherlands (both 6.0%), Belgium (8.7%) and the United Kingdom and Cyprus (both at 9.3%).

6 February 2018

www.euronews.com

Meet the new 'renewable superpowers': nations that boss the materials used for wind and solar

An article from The Conversation

THE CONVERSATION

By Andrew Barron, Sêr Cymru Chair of Low Carbon Energy and Environment, Swansea University

Imagine a world where every country has not only complied with the Paris climate agreement but has moved away from fossil fuels entirely. How would such a change affect global politics?

The 20th century was dominated by coal, oil and natural gas, but a shift to zero-emission energy generation and transport means a new set of elements will become key. Solar energy, for instance, still primarily uses silicon technology, for which the major raw material is the rock quartzite. Lithium represents the key limiting resource for most batteries – while rare earth metals, in particular 'lanthanides' such as neodymium, are required for the magnets in wind turbine generators. Copper is the conductor of choice for wind power, being used in the generator windings, power cables, transformers and inverters.

In considering this future it is necessary to understand who wins and loses by a switch from carbon to silicon, copper, lithium, and rare earth metals.

The countries which dominate the production of fossil fuels will mostly be familiar:

The list of countries that would become the new 'renewables superpowers' contains some familiar names, but also a few wild cards. The largest reserves of quartzite (for silicon production) are found in China, the US, and Russia – but also Brazil and Norway. The US and China are also major sources of copper, although their reserves are decreasing, which has pushed Chile, Peru, Congo and Indonesia to the fore.

Chile also has, by far, the largest reserves of lithium, ahead of China, Argentina and Australia. Factoring in lower-grade 'resources' – which can't yet be extracted – bumps Bolivia and the US onto the list. Finally, rare earth resources are greatest in China, Russia, Brazil – and Vietnam.

Of all the fossil fuel producing countries, it is the US, China, Russia and Canada that could most easily transition to green energy resources. In fact it is ironic that the US, perhaps the country most politically resistant to change, might be the least affected as far as raw materials are concerned. But it is important to note that a completely new set of countries will also find their natural resources are in high demand.

Fossil fuels: largest reserves by country

Oil (billion barrels)	x1	Gas (trillion cubic metres)	x2	Coal (billion tonnes)	x3
Venezuela	301	Iran	34	US	252
Saudi	267	Russia	32	China	244
Canada	172	Qatar	24	Russia	160
Iran	158	Turkmenistan	18	Australia	145
Iraq	153	US	9	India	95

Source: BP Statistical Review of World Energy, June 217

An OPEC for renewables?

The Organization of the Petroleum Exporting Countries (OPEC) is a group of 14 nations that together contain almost half the world's oil production and most of its reserves. It is possible that a related group could be created for the major producers of renewable energy raw materials, shifting power away from the Middle East and towards central Africa and, especially, South America.

This is unlikely to happen peacefully. Control of oilfields was a driver behind many 20th-century conflicts and, going back further, European colonisation was driven by a desire for new sources of food, raw materials, minerals and – later – oil. The switch to renewable energy may cause something similar. As a new group of elements become valuable for turbines, solar panels or batteries, rich countries may ensure they have secure supplies through a new era of colonisation.

China has already started what may be termed 'economic colonisation', setting up major trade agreements to ensure raw material supply. In the past decade it has made a massive investment in African mining, while more recent agreements with countries such as Peru and Chile have spread Beijing's economic influence in South America.

Or a new era of colonisation?

Given this background, two versions of the future can be envisaged. The first possibility is the evolution of a new OPEC-style organisation with the power to control vital resources including silicon, copper, lithium, and lanthanides. The second possibility involves 21st-century colonisation of developing countries, creating super-economies. In both futures there is the possibility that rival nations could cut off access to vital renewable energy resources, just as major oil and gas producers have done in the past.

On the positive side there is a significant difference between fossil fuels and the chemical elements needed for green energy. Oil and gas are consumable commodities. Once a natural gas power station is built, it must have a continuous supply of gas or it stops generating. Similarly, petrol-powered cars require a continued supply of crude oil to keep running.

In contrast, once a wind farm is built, electricity generation is only dependent on the wind (which won't stop blowing any time soon) and there is no continuous need for neodymium for the magnets or copper for the generator windings. In other words solar, wind, and wave power require a one-off purchase in order to ensure long-term secure energy generation.

The shorter lifetime of cars and electronic devices means that there is an ongoing demand for lithium. Improved recycling processes would potentially overcome this continued need. Thus, once the infrastructure is in place, access to coal, oil or gas can be denied, but you can't shut off the sun or wind. It is on this basis that the US Department of Defense sees green energy as key to national security.

A country that creates green energy infrastructure, before political and economic control shifts to a new group of 'world powers', will ensure it is less susceptible to future influence or to being held hostage by a lithium or copper giant. But late adopters will find their strategy comes at a high price. Finally, it will be important for countries with resources not to sell themselves cheaply to the first bidder in the hope of making quick money – because, as the major oil producers will find out over the next decades, nothing lasts forever.

18 February 2018

Running on renewables: how sure can we be about the future?

A variety of models predict the role renewables will play in 2050, but some may be over-optimistic, and should be used with caution, say researchers.

by Hayley Dunning

The proportion of UK energy supplied by renewable energies is increasing every year; in 2017 wind, solar, biomass and hydroelectricity produced as much energy as was needed to power the whole of Britain in 1958.

> **'Research that proposes "optimal" pathways for renewables must be upfront about their limitations if policymakers are to make truly informed decisions.'**
>
> Clara Heuberger

However, how much the proportion will rise by 2050 is an area of great debate. Now, researchers at Imperial College London have urged caution when basing future energy decisions on over-optimistic models that predict that the entire system could be run on renewables by the middle of this century.

Mathematical models are used to provide future estimates by taking into account factors such as the development and adoption of new technologies to predict how much of our energy demand can be met by certain energy mixes in 2050.

These models can then be used to produce 'pathways' that should ensure these targets are met – such as through identifying policies that support certain types of technologies.

However, the models are only as good as the data and underlying physics they are based on, and some might not always reflect 'real-world' challenges. For example, some models do not consider power transmission, energy storage, or system operability requirements.

Reliable supply issues

Now, in a paper published in the journal *Joule*, Imperial researchers have shown that studies that predict whole systems can run on near-100% renewable power by 2050 may be flawed as they do not sufficiently account for reliability of the supply.

Using data for the UK, the team tested a model for 100% power generation using only wind, water and solar (WWS) power by 2050. They found that the lack of firm and dispatchable 'backup' energy systems – such as nuclear or power plants equipped with carbon capture systems – means the power supply would fail often enough that the system would be deemed inoperable.

The team found that even if they added a small amount of backup nuclear and biomass energy, creating a 77%

WWS system, around 9% of the annual UK demand could remain unmet, leading to considerable power outages and economic damage.

Truly informed decisions

Lead author Clara Heuberger, from the Centre for Environmental Policy at Imperial, said: 'Mathematical models that neglect operability issues can mislead decision makers and the public, potentially delaying the actual transition to a low-carbon economy.

> **'If a specific scenario relies on a combination of hypothetical and potentially socially challenging adaptation measures, in addition to disruptive technology breakthroughs, this begins to feel like wishful thinking.'**
>
> Dr Niall Mac Dowell

'Research that proposes 'optimal' pathways for renewables must be upfront about their limitations if policymakers are to make truly informed decisions.'

Co-author Dr Niall Mac Dowell, from the Centre for Environmental Policy at Imperial, said: 'A speedy transition to a decarbonised energy system is vital if the ambitions of the 2015 Paris Agreement are to be realised.

'However, the focus should be on maximising the rate of decarbonisation, rather than the deployment of a particular technology, or focusing exclusively on renewable power. Nuclear, sustainable bioenergy, low-carbon hydrogen, and carbon capture and storage are vital elements of a portfolio of technologies that can deliver this low-carbon future in an economically viable and reliable manner.

'Finally, these system transitions must be socially viable. If a specific scenario relies on a combination of hypothetical and potentially socially challenging adaptation measures, in addition to disruptive technology breakthroughs, this begins to feel like wishful thinking.'

6 March 2018

www.imperial.ac.uk

Predictions – The future of energy

The energy system of the future won't look like today's. The scale of change over the next ten to 20 years will be considerable and, while we don't know exactly what this change will look like, we do know some of the areas that will be important.

One thing seems certain – consumers will play a key role in driving the change as their energy needs for warmth, light, power and, increasingly, mobility change. The energy businesses of the future will provide those services cleanly, cheaply and efficiently by taking advantage of new energy technologies and digital enablers.

The old way of simply sending electrons and gas molecules down the wires and pipes will be replaced by a better, much more sophisticated way of meeting people's needs.

Energy as a service

Energy consumers don't value a kWh (unit) of electricity or a BTU of gas. They value the warmth that it provides, or the light that it enables. That's why energy-as-a-service models are starting to appear where consumers might buy warmth, lighting and power rather than units of electricity and gas. The service would be provided by a business that competes for customers by delivering that warmth, lighting and power most efficiently – perhaps by helping improve home insulation, supplying the best equipment to deliver the required service, sourcing the best energy supplies and optimising any local generation or storage.

With energy as a service, consumers will have more choice in how energy is produced, and the environmental impact of its production, which could lead to more of a focus on locally produced energy. Thanks to the continuing reduction in costs of renewable energy technologies like wind and solar power, we could see the old economies of scale being turned upside down so that generating and using energy locally will represent better value than generating power in relatively few, large, centralised, locations.

Fossil fuels and related technologies have enabled all the major developments in the industrial era, from advances in iron production that made the Industrial Revolution possible, to the advent of the internal combustion engine and its influence on – and benefit from – mass production. Now experts are forecasting that green energy, harnessed with the power of ICT, will drive the next wave of economic development.

Decentralised energy and digital technologies

One of the major benefits of decentralised energy is the move away from large power stations to localised production. That means avoiding the wasted heat in power stations and instead using it locally, and avoiding having to produce the electricity lost in transmission because we have to send it so far. At present, we waste about half of our energy in the UK.

With the rise of decentralised energy, local producer-consumers – or 'prosumers' – will need to flexibly manage the energy in the system, and technologies including energy storage, ways of adjusting usage better (demand-side response) and digital technologies using big data, analytics and cloud computing could help them do that.

For example, there are always surges in energy demand in Britain at half-time during an important sporting event like the football Cup Final or the Wimbledon finals, as everyone gets up and puts the kettle on. To meet this demand, other nonessential household appliances such as fridges or freezers might be run on minimum for just ten minutes to even out the flow of energy without any impact on the food inside them.

The energy market will become largely digital, so that it can integrate all the many parts of the energy world and enable them to work together. The complexity of matching energy demand with energy supply locally and nationally, and integrating storage and demand flexibility, is of such a scale that automation, machine learning and real-time price signals will be needed.

Through the Internet of Things and connected homes concepts though, electronic devices such as washing machines, dishwashers and freezers can all be connected up to use energy at the best price, or stop using energy when there is too much demand.

Free energy?

It's easier than ever for us to generate green energy. The cost of renewable generation equipment is coming down, and there's more capacity on the grid – often to such levels that during sunny days or when there's a lot of wind, there can actually be too much electricity on the grid. The swings that this creates in wholesale prices can mean that prices can go negative, so that there is effectively free energy available. While this can be a problem for managing the grid, the good news is that if consumers can take advantage of these increasing free energy periods by consuming when prices are negative, they can of course save significant sums of money.

Traditionally, consumers have purchased their energy from one of the big six energy suppliers yet, with the rise of microgeneration, people will generate their own power and can sell it back to the grid – so everyone can also be their own energy supplier.

'Vehicle to grid' or V2G could also be a significant solution to the issue of balancing national energy supplies as well as saving money for consumers. Electric car owners could charge them overnight and allow the National Grid to use the stored electricity at times of higher demand the next day.

New technology will make this possible by allowing EV owners to communicate with energy companies via an add-on to their home charging point to ensure they can use their car when they need to, while at times when they don't need the vehicle, its stored energy will contribute a small part of its storage to an intelligent charging system. There is already a scheme in operation for owners of Nissan Leaf vehicles which will earn them cheaper electricity if they sign up.

Getting major energy users on board

The move to new, greener production and consumption of energy will only be successful if major consumers are also on board. It's estimated that the cold chain in the UK currently consumes around 14% of all electricity generated, with food retailers operating massive networks of machines distributed throughout the UK. Tesco Stores Limited is currently participating in a project, funded by Innovate UK, to develop dynamic energy control systems for food retailing refrigeration that could significantly impact on the way the cold chain consumes electricity.

Any changes to refrigerator operational performance must ensure that food safety is not compromised. This project seeks to demonstrate that systems can be developed and delivered, which will enable food retailing refrigeration systems to be linked to demand side response, or DSR, electricity tariffs. Large-scale projects such as this are needed to establish industry interest and confidence in DSR mechanisms, which help the National Grid control demand as well as supply, helping balance the ever-increasing input of renewable generation system.

Consumers at the heart

The biggest change of all, though, is that consumers will go from being on the edge of the energy system to being at its heart. They will have more control over where their energy comes from, how and when they want to consume it, and can take an active role in making sure it doesn't cost the earth.

6 March 2018

www.innovateuk.blog.gov.uk

Nearly 140 countries could be powered entirely by wind, solar and water by 2050

'Our findings suggest that the benefits are so great that we should accelerate the transition to wind, water, and solar, as fast as possible, by retiring fossil-fuel systems early wherever we can.'

By Ian Johnston, Environment Correspondent

More than 70 per cent of the countries in the world – including the UK, US, China and other major economies – could run entirely on energy created by wind, water and solar by 2050, according to a roadmap developed by scientists. And they pointed out that doing so would not only mean the world would avoid dangerous global warming, but also prevent millions of premature deaths a year and create about 24 million more jobs than were lost.

One of the scientists said the social benefits of following their roadmap were so 'enormous' and essentially cost free that human society should 'accelerate the transition to wind, water and solar as fast as possible'. Rooftop solar panels and major solar power plants; offshore and onshore wind turbines; wave, hydroelectric and tidal schemes; and geothermal energy would also be used to replace fossil fuels to generate electricity, power vehicles and heat homes.

The UK is about to publish its own Emissions Reduction Plan, which is supposed to set out how Britain will meet its international commitment in the fight against climate change – to cut emissions by 57 per cent below 1990 levels by 2030.

While the UK has been making good progress on decarbonising electricity generation, the transport and domestic heating sectors remain problematic. As part of its attempts to improve air quality, the Government has announced it will ban the sale of new fossil fuel-powered vehicles in 2040. It remains to be seen how radical it will be in encouraging the switch from gas-central heating to low or zero-carbon methods.

Writing in the journal *Joule*, a team of researchers led by Professor Mark Jacobson, of Stanford University in the US, warned the stakes were high.

'The seriousness of air-pollution, climate, and energy-security problems worldwide requires a massive, virtually immediate transformation of the world's energy infrastructure to 100 per cent clean, renewable energy producing zero emissions,' they said.

'For example, each year, four to seven million people die prematurely and hundreds of millions more become ill from air pollution, causing a massive amount of pain and suffering that can nearly be eliminated by a zero-emission energy system.

'Similarly, avoiding 1.5°c warming since pre-industrial times requires no less than an 80 per cent conversion of the energy infrastructure to zero-emitting energy by 2030 and 100 per cent by 2050.

'Lastly, as fossil-fuel supplies dwindle and their prices rise, economic, social, and political instability may ensue unless a replacement energy infrastructure is developed well ahead of time.'

The roadmaps were developed for 139 countries for which information about energy systems was available, out of the total of 195. They 'describe a future where all energy sectors are electrified or use heat directly with existing technology, energy demand is lower due to several factors, and the electricity is generated with 100% wind, water and sunlight (WWS)', the researchers said.

'The roadmaps are not a prediction of what might happen. They are one proposal for an end-state mix of WWS generators by country and a timeline to get there that we believe can largely solve the world's climate-change, air-pollution, and energy-security problems,' they added.

Professor Jacobson, director of Stanford's atmosphere and energy programme, said political leaders needed reassurance that the transition to a zero-carbon economy would work.

'Both individuals and governments can lead this change. Policymakers don't usually want to commit to doing something unless there is some reasonable science that can show it is possible, and that is what we are trying to do,' he said.

'We are not saying that there is only one way we can do this, but having a scenario gives people direction.'

Fellow researcher Mark Delucchi added: 'It appears we can achieve the enormous social benefits of a zero-emission energy system at essentially no extra cost.

'Our findings suggest that the benefits are so great that we should accelerate the transition to wind, water, and solar, as fast as possible, by retiring fossil-fuel systems early wherever we can.'

The researchers decided to exclude nuclear power, coal with carbon-capture-and-storage, biofuels and bioenergy from their vision of the future.

On nuclear, they highlighted the risks of weapons proliferation and the chance of a power plant meltdown.

'There is no known way at this time to eliminate these risks. By contrast, WWS technologies have none of these risks. Thus, we are proposing and evaluating a system that we believe provides the greatest environmental benefits with the least risk,' the researchers wrote.

23 August 2017

Wind and solar power could provide more than a third of Europe's energy by 2030

By Hayley Dunning

By trading energy between countries with different weather patterns, Europe could make the most of wind and solar power.

This conclusion is from a study modelling the future of weather and energy in Europe, which could help plan future continent-wide energy systems and policies that share renewable resources across countries. The models estimate Europe could use renewables for more than two-thirds of its electricity by 2030, with more than one-third coming from wind and solar.

Wind and solar energy in Europe have quadrupled in use between 2007 and 2016. However, both are susceptible to fluctuating weather patterns, raising concerns about Europe's ability to endure long spells with low winds or overcast skies.

Britain's current heatwave, for example, has led to a 'wind drought', with July 2018 40% less productive than July 2017 for wind power in the UK.

Now, research led by University College Cork in collaboration with Imperial College London and ETH Zürich suggests that despite the unpredictable nature of wind and solar energy, the European power system can comfortably generate at least 35% of its electricity using these renewables alone by 2030, without major impacts on prices or system stability.

30 years of weather data

The study, published today in the journal *Joule*, modelled the impact of long-term weather patterns on wind and solar renewable energy technologies across Europe and their impact on electricity sector operation out to the year 2030.

Previously, researchers have used decades of historic weather data to model this variability in wind and solar energy and its effect on markets, but many studies only analyse data from one given year or focus solely on one country or small region.

In the new study, the team used wind and solar data spanning the 30-year period from 1985 to 2014 to analyse electricity system operation across Europe, including power transmission between countries.

With this dataset, the team was able to model how Europe would fare under five different renewable energy scenarios with varying sustainability ambitions, 12 years into the future.

Pooling European resources

Dr Seán Collins from University College Cork, a co-author of the study, said: 'When planning future power systems with higher levels of wind and solar generation, one year of weather data analysis is not sufficient. We find that single-year studies could yield results that deviate by as much as 9% from the long-term average at a European level and even more at a country level.'

Co-author Dr Iain Staffell, from the Centre for Environmental Policy at Imperial, added: 'If European countries pool their renewable energy resources, everyone benefits from a cheaper and more secure electricity system, which will ultimately allow wind and solar power to provide a greater share of our energy needs.'

By using multiple years the team found that in future scenarios, carbon dioxide emissions and the total costs of generating energy could fluctuate wildly, becoming up to five times more uncertain as weather-dependent resources gain greater traction in the market.

However, they also found that Europe could withstand this variability quite well thanks to its close integration, for example by trading energy between regions and countries with different weather patterns.

Cheaper renewable energy

The team believe their models and data could be used to depict a variety of possible future scenarios to help policymakers better understand the reliability and impact of renewable energy, including the impacts of a shift to 100% renewable electricity systems.

Dr Staffell added: 'The world, and especially Europe, is rapidly shifting towards renewable energy sources, so it is important that we develop new tools to help understand the impacts this will have on the day-to-day running of the power system and how much households can expect to pay for clean electricity.

'While the cost of electricity might vary more from one year to the next along with the weather, overall the costs are going down as renewable sources become the cheapest way to produce electricity.'

By making their models and data openly available, the researchers also hope that future work will continue with greater awareness of these long-term weather patterns to accurately depict a more renewable energy-reliant world.

26 July 2018

www.imperial.ac.uk

Fairtrade renewable energy: shedding light on clean energy's dirty secrets

An article from The Conversation

THE CONVERSATION

By David Flynn, Professor, Embedded Intelligence in Energy Systems, Heriot-Watt University, Merlinda Andoni, Research Associate, Heriot-Watt University, and Valentin Robu, Associate Professor, Smart Systems Group, Heriot-Watt University

The world is coming together on renewable energy. Trust in technology such as solar and wind power generation is increasingly reflected in the investment it receives from governments around the globe. In the UK, fairness to the consumer and affordability are at centre stage as the country look to eradicate fuel poverty on the road to a zero-carbon future.

The Scottish government is considering a not-for-profit energy utility to compensate the consumer by reducing the amount of energy trading in the private sector. Elsewhere, an analysis of the value of smart meters – the household devices which track energy usage and charge consumers accordingly – returned an unsatisfactory verdict. The saving for the UK consumer was found to be only £11 per year – disappointing in light of the £11 billion investment for the switchover to smart meters.

In a similar vein of 'fair' energy, eliminating the sector's carbon footprint is the other focus of this debate, particularly in discussions about 'climate-conscious investment'. The Bishop of Oxford, Steven Croft, recently used the idea to urge investors 'to divest from any fossil fuel company which is not on an unequivocal path by 2020 to aligning its business investment plan with the Paris Agreement to restrict global warming to well below 2°C'.

But there's a reality we're all overlooking when it comes to the future of our energy supply. That is, we're only applying a sense of fairness to our own society. We must take a global view of fair trade in energy, which is conscious of the true cost of our low to zero-carbon technologies.

The true cost of power

In June this year it was reported that child labour was being used in dangerous cobalt mining in the Democratic Republic of Congo (DRC). The mineral cobalt is used in virtually all batteries of devices we use everyday, including mobile phones and computers – but also electric vehicles. A report by Amnesty International first revealed that cobalt mined by children was ending up in products from several companies, including Apple, Microsoft, Tesla and Samsung. The latest research by the United Nations Children's Fund (UNICEF) estimates 40,000 children are working in DRC mines.

Cobalt plays a significant role in the development of many prominent renewable energy technologies such as solar power, wind power and biogas. Although cobalt is not directly used in the solar panels themselves, rechargeable batteries containing cobalt are important to store the energy produced.

Cobalt can be found in the cathode of lithium-ion batteries. In wind turbines, permanent magnets are used in generators to create a magnetic field without an input of electricity, allowing the turbines to run at lower wind speeds while still producing energy. The use of permanent magnets also negates the need for a gearbox in the wind turbine, improving reliability and lowering maintenance costs. While a significant engineering achievement, cobalt sourced from child labour is an important component here.

Cobalt is also a bio-essential element, which means organisms need it for growth and reproduction and it plays a key role in renewable biogas technology. Biogas is a type of biofuel that is naturally produced from the decomposition of organic waste.

Fairtrade food – why not energy?

So why hasn't the sourcing of cobalt provoked a global reaction akin to that which launched Fairtrade clothing and food? Is energy, which underpins all of our infrastructure and services, exempt from responsible trading? Are governments willing to turn a blind eye to the ethical issues of a low-carbon future?

Our transition to a low-carbon economy is not a movement that will lose pace – in fact the European Commission estimates that a €180 billion annual investment in clean energy is required to keep the rise in global temperature below 2°C. So we have to ask the question – are we exacerbating exploitation as we invest in low-carbon technologies without due diligence to the ethics involved?

To avoid public money funding child exploitation, we should apply a Fairtrade certification scheme to technology that promotes better prices, decent working conditions,

local sustainability and fair terms for workers in developing countries.

Fair energy of the future

As we wrestle with the challenges of a sustainable energy future, it's clear we need to take a broader view which embodies the interests of all people, from the cobalt miners to the energy consumers. It should also maximise the benefits of multiple technologies.

Researchers within the Smart Systems Group of Heriot-Watt University, are exploring how energy security, sustainability and fairness can be achieved through binding these various technologies into one system. Blockchain technology, a digital log which can track transactions in money and services between users, has been successfully used to curb the trade in blood diamonds by imposing strict standards on where the diamonds come from. In energy, it could be used to trace both the source of the electricity consumers buy, and the source of the materials used in making batteries as well.

Ideas such as the Circular Economy are challenging manufacturers to reuse and recycle materials. Businesses which currently rely on cobalt in batteries and wind turbines can create new revenue streams by moving away from selling batteries to single customers, to a business model similar to hire purchase where a battery can be shared by several parties as needed. This would reduce the demand for cobalt and new batteries and provide a more efficient distribution of energy.

There are abundant technical tools and financial incentives which can help governments enforce Fairtrade in the energy sector. While we strive for an energy system which is conscious of the threat of climate change, we must not lose our social conscience and ignore where the materials to build that system come from.

3 August 2018

Fatbergs clogging up Britain's sewers could soon be providing homes with green energy

From dangerous sewage to useful methane.

By Alex Finnis

Fatbergs clogging up sewers in Britain's cities could soon be providing homes with green energy, according to new research. A technique to break down the solid masses of congealed fat, wet wipes, nappies, oil and condoms has been developed by scientists. The flushed waste can grow into fatbergs like the 130-ton, 250-metre-long monster that blocked up a Victorian tunnel in Whitechapel, east London, a year ago.

Earlier this year a Channel 4 documentary unveiled another three times as long – under the capital's South Bank.

From dangerous sewage to green energy

But the fats, oil and grease, known collectively as FOG, could now potentially do some environmental good by being turned into green biogas methane. In experiments, a team of Canadian scientists heated them to between 90 and 110 degrees centigrade. Adding hydrogen peroxide – a chemical that kickstarts the breakdown of organic matter – then reduced the volume of solids by up to 80 per cent. It also released fatty acids from the mixture that can be broken down by bacteria in the next stage of treatment.

Methane gas is a valuable renewable energy source

Engineer Dr Asha Srinivasan from the University of British Columbia explained: 'FOG is a terrific source of organic material that microorganisms can feed on to produce methane gas, which is a valuable, renewable energy source. 'But if it is too rich in organics, bacteria cannot handle it and the process breaks down.

'By preheating it to the right temperature, we ensure that the FOG is ready for the final treatment and can make the maximum amount of methane.'

Her team's method will enable farmers to load more FOG into their biogas digesters – the large tanks that treat farm wastes, including cow manure, to produce methane. Dr Srinivasan said: 'Farmers typically restrict FOG to less than 30 per cent of the overall feed.

'But now the FOG can be broken down into simpler forms, so you can use much more than that, up to 75 per cent of the overall feed.

'You would recycle more oil waste and produce more methane at the same time.'

FOG could eventually help power our cities

Lead researcher Professor Victor Lo said, ultimately, the technology can be used in municipal FOG management programmes. He added: 'The principle would be the same. You pretreat the FOG so it doesn't clog the pipes, and add it to sewage sludge to produce methane from the mix.

'To the best of our knowledge, this type of pretreatment for FOG has not been studied before, although simple chemical methods do exist to break down FOG.

'We are hoping to do more research to find the optimal ratio of FOG to dairy manure so they can be pretreated together.'

750-metre fatberg was found under London last year

Fatbergs form into huge concrete-like slabs and can be found beneath almost every UK city, growing larger with every flush. They also include food wrappers and human waste, blocking tunnels and raising the risk of sewage flooding into homes. They can grow metres tall and hundreds of metres long, with water providers declaring an epidemic of fatberg emergencies in 23 UK cities, costing tens of millions of pounds to get rid of.

The biggest ever discovered in the UK featured in Channel 4's *Fatberg Autopsy: Secrets of the Sewers*, when cameras followed a team of eight 'flushers' as they chipped away at the 750-metre whopper under the South Bank. The one discovered beneath Whitechapel last September was longer than Tower Bridge – and weighed as much as 11 double-decker buses. Fatbergs take weeks to remove and form when people put things they shouldn't down sinks and toilets.

The study was published in the journal Water, Air, & Soil Pollution.

3 August 2018

Poo power: is faecal matter the future?

With fossil fuels running low and the human population increasing by the second, the search is on to find the most economically and environmentally sufficient source of energy. The limited amount of space for waste and landfill within the UK means that it is now time to focus on the recycling of waste, rather than the storage of it.

Meanwhile, the UK produces a massive 1.73 million tons of sewage every year… I think you see where we're going. Read on to see how sewage can be used for power.

How can poo produce power?

With so much waste being produced, there's clearly huge potential to generate a substantial amount of energy if handled properly. And the use of animal and human waste as a source of energy is nothing new. The Chinese have been using covered sewage tanks dating back to the 13th century as a source of energy, and Bombay leper colonies in 1859 were pumping sewage through anaerobic digestion plants to create power.

The process isn't as brain-numbing nor stomach-churning as you may think. Once the sewage has been separated into waste and clean water, anaerobic digesters break down the leftover waste into what can be described as odourless sludge and methane. Analysis of large biological samples is one of the topics covered in the article 'Ocean-Going Lab… The Real Test of a Nutrient Analyser'. The waste solids are passed on to create fuel and fertiliser while the methane is passed through a biogas plant where impurities are removed and a 'gas-like' smell is added. This biomethane is then fed back into the national network.

The reality of the power of poo

The idea of using human faeces as a source of energy in the UK is becoming more of a reality as the 2020 target of 15% of energy consumption to come from renewable sources is rapidly approaching. Both Northumbrian and Yorkshire Water have made it their mission to recycle their sewage in the form of power with an £8 million gas to grid plant in Howden and a £72 million investment project under way in Leeds.

The poo-power treatment facility under construction in Leeds promises the ability to convert 131 tonnes of sludge into power each day, reducing the site's carbon emission by 15% and providing enough power for 8,000 homes. Although the hefty price tag, over time the power of poo may be enough to not only cut down the UK's carbon footprint, but also create a multimillion pound industry.

So, in a world run by telecommunication and pharmaceutical stocks and shares, would you invest in poo?

10 January 2018

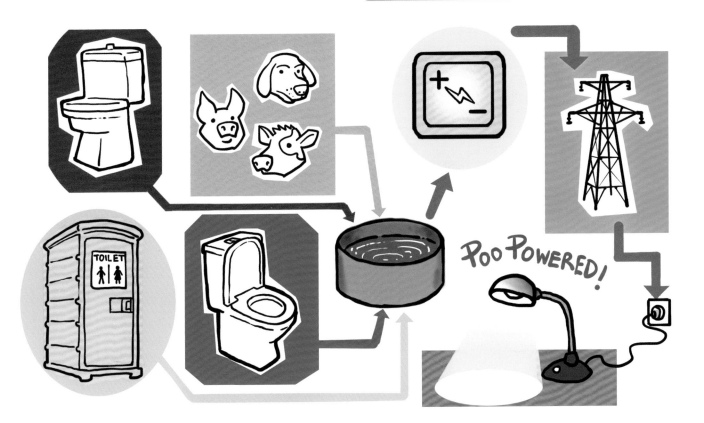

How we can turn plastic waste into green energy

An article from The Conversation.

THE CONVERSATION

By Anh Phan, Lecturer in Chemical Engineering, Newcastle University

In the adventure classic *Back to the Future*, Emmett 'Doc' Brown uses energy generated from rubbish to power his DeLorean time machine. But while a time machine may still be some way off, the prospect of using rubbish for fuel isn't too far from reality. Plastics, in particular, contain mainly carbon and hydrogen, with similar energy content to conventional fuels such as diesel.

Plastics are among the most valuable waste materials – although with the way people discard them, you probably wouldn't know it. It's possible to convert all plastics directly into useful forms of energy and chemicals for industry, using a process called 'cold plasma pyrolysis'.

Pyrolysis is a method of heating, which decomposes organic materials at temperatures between 400°C and 650°C, in an environment with limited oxygen. Pyrolysis is normally used to generate energy in the form of heat, electricity or fuels, but it could be even more beneficial if cold plasma was incorporated into the process, to help recover other chemicals and materials.

The case for cold plasma pyrolysis

Cold plasma pyrolysis makes it possible to convert waste plastics into hydrogen, methane and ethylene. Both hydrogen and methane can be used as clean fuels, since they only produce minimal amounts of harmful compounds such as soot, unburnt hydrocarbons and carbon dioxide (CO_2). And ethylene is the basic building block of most plastics used around the world today.

As it stands, 40% of waste plastic products in the US and 31% in the EU are sent to landfill. Plastic waste also makes up 10% to 13% of municipal solid waste. This wastage has huge detrimental impacts on oceans and other ecosystems.

Of course, burning plastics to generate energy is normally far better than wasting them. But burning does not recover materials for reuse, and if the conditions are not tightly controlled, it can have detrimental effects on the environment such as air pollution.

In a circular economy – where waste is recycled into new products, rather than being thrown away – technologies that give new life to waste plastics could transform the problem of mounting waste plastic. Rather than wasting plastics, cold plasma pyrolysis can be used to recover valuable materials, which can be sent directly back into industry.

How to recover waste plastic

In our recent study we tested the effectiveness of cold plasma pyrolysis using plastic bags, milk and bleach bottles collected by a local recycling facility in Newcastle, UK.

We found that 55 times more ethylene was recovered from [high density polyethylene (HDPE)] – which is used to produce everyday objects such as plastic bottles and piping – using cold plasma, compared to conventional pyrolysis. About 24% of plastic weight was converted from HDPE directly into valuable products.

Plasma technologies have been used to deal with hazardous waste in the past, but the process occurs at very high temperatures of more than 3,000°C, and therefore requires a complex and energy intensive cooling system. The process for cold plasma pyrolysis that we investigated operates at just 500°C to 600°C by combining conventional heating and cold plasma, which means the process requires relatively much less energy.

The cold plasma, which is used to break chemical bonds, initiate and excite reactions, is generated from two electrodes separated by one or two insulating barriers.

Cold plasma is unique because it mainly produces hot (highly energetic) electrons – these particles are great for breaking down the chemical bonds of plastics. Electricity for generating the cold plasma could be sourced from renewables, with the chemical products derived from the process used as a form of energy storage: where the energy is kept in a different form to be used later.

The advantages of using cold plasma over conventional pyrolysis is that the process can be tightly controlled, making it easier to crack the chemical bonds in HDPE that effectively turn heavy hydrocarbons from plastics into lighter ones. You can use the plasma to convert plastics into other materials; hydrogen and methane for energy, or ethylene and hydrocarbons for polymers or other chemical processes.

Best of all, the reaction time with cold plasma takes seconds, which makes the process rapid and potentially cheap. So, cold plasma pyrolysis could offer a range of business opportunities to turn something we currently waste into a valuable product.

The UK is currently struggling to meet a 50% household recycling target for 2020. But our research demonstrates a possible place for plastics in a circular economy. With cold plasma pyrolysis, it may yet be possible to realise the true value of plastic waste – and turn it into something clean and useful.

1 October 2018

Promoting a sustainable energy future

Facts and figures on the UK power mix, environmental programmes and offshore transmission

What generates Britain's power?

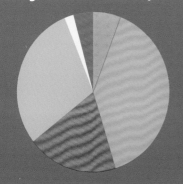

- Net imports 4% (15 twh)
- Other sources 2% (6 twh)
- Renewable sources 29% (95 twh)
- Nuclear plant 19% (64 twh)
- Gas-fired turbines 39% (131 twh)
- Coal-fired power stations 6% (21 twh)

Reducing the UK's carbon emissions and increasing the energy we get from sources like renewables, will help to ensure the security and sustainability of our energy supplies for generations to come. With the protection of future and existing consumers' interests at our core, we run a range of government consumer and environmental programmes and manage the regulatory regime for offshore electricity transmission networks.

Environmental programmes

£7 Billion The value of environmental schemes we've administered during 2016/17.

Helping people in difficult situations

£1.78 BILLION Amount of support to fuel poor households through the Warm Home Discount since its launch in 2011.

£323 MILLION How much suppliers paid in support to people who needed it in 2016/17.

The UK government has announced the WHD scheme will continue to support fuel poor customers until 2020/21 at current levels of £320 million a year, rising with inflation.

Promoting energy efficiency and renewable electricity

2,439,624 The number of efficiency measures installed under the Energy Company Obligation across over *1.9 million households* to December 2018.

Certificates (ROCs) issued for eligible renewable electricity generated under the Renewables Obligation in 2017/19. **100,902,554**

828,450 The number of solar panel installations registered for the Feed-in Tariff to 5 January 2019.

Generating heat through more renewable sources

65,906 Domestic Renewable Heat Incentive Number of renewable heat installations accredited to 5 January 2019

4.4GW Non-Domestic Renewable Heat Incentive The total installed capacity to 31 December 2018

Offshore transmission

£5 BILLION Investment value in 6.9GW of offshore transmission projects under our competitive tender process (Tender Rounds 1-5).

£700 MILLION Estimated savings for consumers resulting from our competitive tender process to connect offshore wind farms to the GB high voltage grid.
(CEPA and BDO evaluation Offshore transmission Review based on Net Present Value figures at 2014/15 price levels)

Source: Ofgem
UK electricity supply mix data from Department for Business, Energy and Industrial Strategy: Energy Trends: Electricity (March 2018).
Environmental programmes data from Ofgem public reports. Offshore estimated savings from Department of Energy and Climate Change for Tender Rounds 1-3.
Information correct at January 2019.
Find out more at www.ofgem.gov.uk/facts-and-figures

Energy saving quick wins

Whether you're a homeowner, private or social renter, student, or you live at home with your parents, there are many things you can do to reduce how much energy you use and how much is spent.

We're all responsible for the energy we use in our homes. Take a look at our quick tips and see if you're saving as much energy as you could be.

1. Understand your bill

The information on a typical energy bill can be confusing, but understanding it can go a long way to helping you get to grips with your energy usage at home.

2. Switch off standby

You can save around **£30 a year** just by remembering to turn your appliances off standby mode.

Almost all electrical and electronic appliances can be turned off at the plug without upsetting their programming. You may want to think about getting a standby saver which allows you to turn all your appliances off standby in one go.

Check the instructions for any appliances you aren't sure about. Some satellite and digital TV recorders may need to be left plugged in so they can keep track of any programmes you want to record.

3. Careful in your kitchen

You can save around **£36 a year** from your energy bill just by using your kitchen appliances more carefully:

⇨ Use a bowl to wash up rather than a running tap and save **£25 a year** in energy bills.

⇨ Cutback your washing machine use by just one cycle per week and save **£5 a year** on energy.

⇨ Only fill the kettle with the amount of water that you need and save around **£6 a year**.

4. Get a head

If you've got a shower that takes hot water straight from your boiler or hot water tank (rather than an electric shower), fit a water-efficient shower head. This will reduce your hot water usage while retaining the sensation of a powerful shower.

A water-efficient shower head could save a four-person household (e.g. a family of four or even a shared student flat) as much as **£70 a year** on gas for water heating, as well as a further **£120 a year** on water bills if they have a water meter.

How much could you save?
Add up the potential £££ savings for each energy saving action

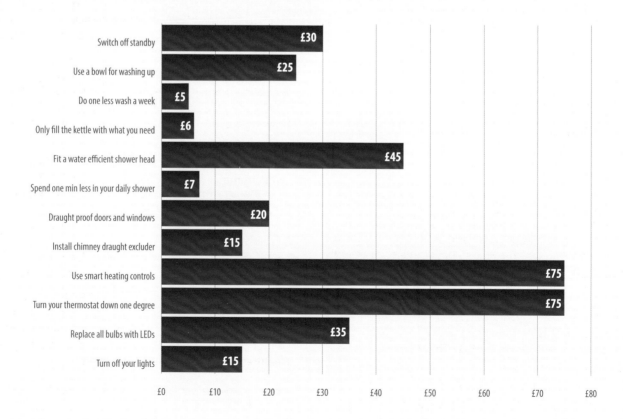

Source: Energy Saving Trust

Please note, all savings are based on figures for Great Britain and are not applicable to Northern Ireland.

Calculation is based on the assumption that a family of 4 takes 20 showers a week and replaces a 13 litre/minute power-shower head with a 7.7 litre/min water-efficient shower head, and the family are charged £2.97 per cubic meter of water used (includes sewage charge).

5. Spend less time in the shower

Spending one minute less in the shower each day will save up to **£7 a year** off your energy bills, per person. With a water meter this could save a further **£12** off annual water and sewerage bills.

If everyone in a four-person household did this it would lead to a total saving of **£75 a year**.

6. Draught proofing

Unless your home is very new, you will lose some heat through draughts around doors and windows, gaps around the floor, or through the chimney.

Professional draught-proofing of windows, doors and blocking cracks in floors and skirting boards can cost around £200, but can save around **£20 a year** on energy bills. DIY draught proofing can be much cheaper.

Installing a chimney draught excluder could save around **£15 a year** as well.

7. Take control of your heating

More than half the money spent on fuel bills goes towards providing heating and hot water.

Installing a room thermostat, a programmer and thermostatic radiator valves and using these controls efficiently could save you around £75 a year.

If you already have a full set of controls, turning down your room thermostat by just one degree can save around £75 a year.

Whatever the age of your boiler, the right controls will allow you to:

⇨ Set your heating and hot water to come on and off when you need them

⇨ Heat only the areas of your home that need heating

⇨ Set the temperature for each area of your home.

8. Get savvy with smart controls

Smart heating controls are the latest innovation to help you control your heating and understand your energy use.

They allow you to control your heating remotely via a mobile app, meaning that you can manage the temperature of your home from wherever you are, at whatever time of day.

9. Switch to LEDs

You can now get LED spotlights that are bright enough to replace halogens, as well as regular energy saving bulbs ('compact fluorescent lamps' or CFLs). They come in a variety of shapes, sizes and fittings.

If the average household replaced all of their bulbs with LEDs, it would cost about £100 and save about **£35 a year** on bills.

Are you **really** sure you want that much water for only **one** cup?

It's going to cost you!

SMART APPLIANCES.

10. Turn off lights

Turn your lights off when you're not using them. If you switch a light off for just a few seconds, you will save more energy than it takes for the light to start up again, regardless of the type of light.

This will save you around **£15 a year** on your annual energy bills.

Are you a homeowner?

If you're a homeowner, there are some other things you can consider to improve the energy efficiency of your home. These can be more costly to put in place, but will benefit you in the long term.

⇨ **Explore the benefits of renewable energy technology.** Installing renewable technology for your heating or electricity can lead to greater savings on your energy bills and extra income as a result of the energy you generate.

⇨ **Consider your options for insulation.** Making sure your home is well insulated can significantly reduce unnecessary heat loss – leading to lower energy bills and a more comfortable home.

Are you renting?

Potential tenants don't always apply the same level of scrutiny to their potential home as buyers. But when it comes to energy efficiency, this could be a key mistake.

⇨ **Think energy efficiency when renting a property.** Comfort and keeping bills under control are key for renters, which is why it is important to check energy features of potential rental properties.

www.energysavingtrust.org.uk

Key facts

- In 2016, global renewable energy capacity grew by a record amount while its cost fell considerably. (page 3)

- 161GW of renewable energy capacity was installed worldwide in 2016. (page 3)

- 2017 was declared the UK's 'greenest year ever' by WWF. (page 6)

- The National Grid said the best day for wind generation was November 28 which produced 116,599 megawatts (MW) – enough to power 9.59 million homes. (page 8)

- Over the course of a year, about 15% of the UK's electricity comes from wind power. (page 10)

- Studies show that so-called Blue Energy has the potential of meeting 10% of the EU's power demand by 2050. (page 11)

- The legislative package 'Clean Energy Package for all Europeans', sets a renewable energy target for the EU by 2030 of 32%. (page 11)

- Hydropower makes up nearly 71 per cent of the world's renewable energy. (page 12)

- 62% of people said they wanted to fit solar. (page 13)

- 71% would join a local energy scheme such as a community windfarm or solar panel collective. (page 13)

- 68% thought the big six energy suppliers' market dominance should be broken up to allow smaller clean energy firms to grow. (page 13)

- The UK gets around 60% of the solar radiation found in the Equator. (page 14)

- There were around 230,000 solar power projects in the UK by the end of 2011. (page 14)

- Using wind and solar energy in place of fossil fuels helped avoid between 3,000 and 12,700 premature deaths in the US between 2007 and 2015, saving US$35–220 billion. (page 19)

- In 2014, the amount of renewable energy share reached 14% of total energy consumption and was estimated to reach 16.4% in 2015. (page 20)

- In 2011, renewables created 21.7% of the EU's power. (page 20)

- The costs of photovoltaic modules fell by 80% between the end of 2009 and late 2015. (page 20)

- EU countries hold 30% of the patents in renewable energy globally. (page 21)

- It has been suggested 2.2 jobs were lost for every job that the green energy industry created. (page 22)

- In January 2015 Spain set a new daytime record, when wind energy accounted for 54 per cent of their total energy used. (page 22)

- Sweden gets more than half of all its energy needs from renewables. (page 23)

- By 2020, 15 per cent of Poland's energy needs should come from renewables. (page 23)

- Europe has doubled its renewable use in the past 12 years. (page 24)

- The energy we waste in Europe could power all our buildings' needs. (page 24)

- Energy efficient products could save families up to 500 Euros a year. (page 24)

- More than a quarter of the innovations for new tech in renewables are made by European Companies. (page 24)

- The Ocean at Europe's doorstep could eventually power 10% of all our demand. (page 24)

- In 2016, Eurostat figures showed that Austria (72.6%) and Sweden (64.9%) produced at least three fifths of all their electricity from renewable energy sources, while Portugal (54.1%), Denmark (53.7%) and Latvia (51.3%) produced more than half. (page 24)

- Imperial researchers have shown that studies that predict whole systems can run on near-100% renewable power by 2050. (page 27)

- More than 70 per cent of the countries in the world could run entirely on energy created by wind, water and solar by 2050. (page 30)

- Wind and solar energy in Europe have quadrupled in use between 2007 and 2016. (page 31)

- July 2018 was 40% less productive than July 2017 for wind power in the UK. (page 31)

- An analysis of the value of smart meters returned an unsatisfactory verdict. The saving for the UK consumer was found to be only £11 per year – disappointing in light of the £11 billion investment for the switchover. (page 32)

- The latest research by UNICEF estimates 40,000 children are working in DRC mines. (page 32)

- The European Commission estimates that a €180 billion annual investment in clean energy is required to keep the rise in global temperature below 2°C. (page 33)

- The UK produces a massive 1.73 million tons of sewage every year. (page 35)

- Bombay leper colonies in 1859 were pumping sewage through anaerobic digestion plants to create power. (page 35)

- Poo-power treatment facility under construction in Leeds promises the ability to convert 131 tonnes of sludge into power each day, reducing the site's carbon emission by 15% and providing enough power for 8,000 homes. (page 35)

- As it stands, 40% of waste plastic products in the US and 31% in the EU are sent to landfill. (page 36)

- The UK is currently struggling to meet a 50% household recycling target for 2020. (page 36)

Biofuel

A gaseous, liquid or solid fuel derived from a biological source, e.g. ethanol, rapeseed oil. Some scientists claim that GM would be a useful tool in the quest to produce biofuels which would be beneficial for the environment.

Biofuels and biomass

Plants use photosynthesis to store energy from the Sun in their leaves and stems. Living things, like these plant materials, are known as biomass. The wide range of fuels derived from biomass are known as biofuels. Corn ethanol, sugar ethanol and biodiesel are the primary biofuels markets.

Energy

A force which powers or drives something. It is usually generated by burning a fuel such as coal or oil, or by harnessing natural heat or movement (for example, by using a wind turbine).

Fossil fuels

Fossil fuels are stores of energy formed from the remains of plants and animals that were alive millions of years ago. Coal, oil and gas are examples of fossil fuels. They are also known as non-renewable sources of energy, because they will eventually be used up: as they are finite, once they are gone we will be unable to produce more of them.

Geothermal power

The Earth is hot inside. Most of this heat comes from radioactive decay, which heats up the surrounding rocks. This heat can be used as an energy source: water is pumped down into the hot rocks, and the steam produced used to drive an electricity generator. The hot water can also be used directly to heat homes and businesses. However, unless the rock conditions are just right, geothermal power is not cost-effective.

Green energy

The same as renewable energy, which come from natural resources rather than non-renewable sources. It's called 'green' due to the fact that the sources are environmentally friendly, sustainable and have zero emissions.

Hydropower (or water power)

Energy which is generated using the movement of running water. This includes tidal/wave power.

Microgeneration

Microgeneration is the production of heat or electricity by individual households and small businesses. Microgeneration technologies are low- or even zero-carbon and allow householders and business owners to generate their own sustainable heat and/or electricity

Non-renewable energy

Energy generated from finite resources such as fossil fuels, energy can be generated from these sources however, they will eventually run out.

Nuclear power

A method of generating energy using controlled nuclear reactions. These are used to create steam, which then powers a generator. Nuclear power is controversial and subject to much debate, with proponents saying it is a greener and more sustainable alternative to fossil fuels, whereas opponents argue that nuclear waste is potentially hazardous to people and the environment.

Offshore wind farm

An offshore wind farm consists of a number of wind turbines, constructed in an area of water where wind speeds are high in order to maximise the amount of energy which can be generated from wind.

Renewable energy

Energy generated from natural resources such as wind, water or the Sun. Unlike fossil fuels, energy can be generated from these sources indefinitely as they will never run out.

Solar power

Energy generated by harnessing the heat of the Sun.

Wind power

Energy which is generated using movement powered by wind. This is most commonly achieved via wind turbines, which are used to produce electricity.

Assignments

Brainstorming

⇨ In small groups, discuss what you know about energy and energy alternatives. Consider the following points:

- What is renewable energy?

- What is non-renewable energy?

- What does it mean to be 'energy efficient'?

⇨ Create a mind map of all the different types of energy that you can think of. Use different colours for renewable and non-renewable energies.

⇨ In groups, each person writes down their ideas for new renewable energy sources. Then they pass it along to the next person and they write how the idea could be expanded. Do this until each person has contributed, then discuss the options on each piece of paper.

Research

⇨ Research possible sources of renewable energy that could be used instead of fossil fuels in the future. In your opinion, which is the most viable energy source? Write a short summary of your findings and conclusions based on the research you have carried out.

⇨ Create a short questionnaire to find out people's attitudes to different types of energy. Try to ask at least 20 people. Consider the differences in attitudes between the different age groups.

⇨ Research renewable energy projects in your area. Does your school or college use renewable energy? Create a short report on your findings.

⇨ Choose a country in the EU and research their use of renewable energy.

Design

⇨ Design a poster to help people save energy. Use the article *Energy saving quick wins* as a guide.

⇨ Create a leaflet that will inform UK homes about renewable energy and its potential advantages and disadvantages.

⇨ Choose one of the articles in this book and create an illustration to highlight the key themes/message of your chosen article.

⇨ Create an infogram using the article *Eight awesome facts about renewable energy*.

⇨ Choose one type of renewable energy and create a poster to promote it.

Oral

⇨ Imagine a proposal has been put forward for a solar farm or a wind farm to be built on your school playing field. Role play a local council meeting, with one group of students acting as the councillors supporting the proposal, another group of students acting as some local residents who oppose the proposal, and another group of students acting as some local residents who support either the solar farm or the wind farm. Could the councillors be accused of not taking residents' concerns into account? What might be their concerns? Which would be the best option solar or wind power?

⇨ Research different energy suppliers in the UK and create a five-minute presentation. What kind of different tariffs do they offer , do they offer fixed price energy? How do they compare? Do they offer renewable energy options?

⇨ As a class, discuss the following statement: 'Poo power: is faecal matter the future of energy?'

⇨ In pairs, discuss the pros and cons of micro-generation of renewable energy.

Reading/writing

⇨ Read the article *Types and alternative sources of renewable energy* and write a summary for your school newsletter.

⇨ Write an article exploring renewable energy in Europe. Which countries use the most renewable energy? Which uses lots of types of energy?

⇨ Write a letter to your headteacher to persuade them to use renewable energy in your school. Which types do you think will be the best for your school? Use persuasive language to get your point across.

⇨ Write a short story imaging what the world would be like if we only had renewable energy. Do you think it would be much different that it is now? Or, do you think that it would be drastically different?

⇨ Choose an article in this book and write a short summary. Pick out five key points that you feel are important.

Acknowledgements

The publisher is grateful for permission to reproduce the material in this book. While every care has been taken to trace and acknowledge copyright, the publisher tenders its apology for any accidental infringement or where copyright has proved untraceable. The publisher would be pleased to come to a suitable arrangement in any such case with the rightful owner.

Images

All images courtesy of iStock except pages 6, 7, 9, 11, 12, 16, 18, 19, 20, 21& 32 Unsplash. Page 1 Rawpixel and pages 3, 33, 34 & 36 Pixabay.

Icons

Icons on pages 2, 20, 24 & 37 were made by Freepik from www.flaticon.com.

Icons on page 14 were made by Pixabay.

Illustrations

Don Hatcher: pages 8 & 17. Simon Kneebone: pages 19 & 28. Angelo Madrid: pages 35 & 39.

Additional acknowledgements

With thanks to the Independence team: Shelley Baldry, Tina Brand, Danielle Lobban, Jackie Staines and Jan Sunderland.

Tracy Biram

Cambridge, January 2019